METHODS FOR ASSESSING THE STRUCTURAL RELIABILITY OF BRITTLE MATERIALS

A symposium
sponsored by
ASTM Committee E-24
on Fracture Testing
San Francisco, Calif., 13 Dec. 1982

ASTM SPECIAL TECHNICAL PUBLICATION 844
Stephen W. Freiman, National Bureau of
Standards, and C. Michael Hudson,
NASA Langley Research Center, editors

ASTM Publication Code Number (PCN)
04-844000-30

 1916 Race Street, Philadelphia, Pa. 19103

Library of Congress Cataloging in Publication Data

Methods for assessing the structural reliability of
brittle materials.

(ASTM special technical publication; 844)
"ASTM publication code number (PCN) 04-844000-30."
Includes bibliographies and index.
1. Fracture mechanics—Congresses. 2. Brittleness—
Congresses. 3. Ceramic materials—Congresses. I. Frei-
man, S. W. II. Hudson, C. M. III. ASTM Committee E-24
on Fracture Testing. IV. Series.
TA409.M46 1984   620.1'126   83-73253
ISBN 0-8031-0265-8

NOTE

The Society is not responsible, as a body,
for the statements and opinions
advanced in this publication.

Printed in Baltimore, Md. (b)
October 1984

Foreword

The symposium on Methods for Assessing the Structural Reliability of Brittle Materials was held on 13 Dec. 1982 in San Francisco, Calif. The event was sponsored by ASTM Committee E-24 on Fracture Testing. Stephen W. Freiman, National Bureau of Standards, and C. Michael Hudson, NASA Langley Research Center, presided as chairmen of the symposium and also served as editors of this publication.

Related
ASTM Publications

Fractography of Ceramic and Metal Failures, STP 827 (1984), 04-827000-30

Fracture Mechanics for Ceramics, Rocks, and Concrete, STP 745 (1981). 04-745000-30

Fractography and Materials Science, STP 733 (1981), 04-733000-30

Fracture Mechanics Applied to Brittle Materials (11th Conference), STP 678 (1979), 04-678000-30

A Note of Appreciation
to Reviewers

The quality of the papers that appear in this publication reflects not only the obvious efforts of the authors but also the unheralded, though essential, work of the reviewers. On behalf of ASTM we acknowledge with appreciation their dedication to high professional standards and their sacrifice of time and effort.

ASTM Committee on Publications

ASTM Editorial Staff

Contents

Introduction

How can we ensure that ceramic components designed for gas turbine engines, human prostheses, optical communication lines, and many other varied applications will survive the in-service stresses imposed on them? This symposium on Methods for Assessing the Structural Reliability of Brittle Materials was organized under the auspices of two subcommittees of ASTM Committee E-24 on Fracture Testing—Subcommittee E24.06 on Fracture Mechanics Applications and Subcommittee E24.07 on Fracture Toughness of Brittle Nonmetallic Materials—for the purpose of providing a forum for discussion of current and proposed procedures for using fracture mechanics data in the design of structures made from essentially brittle materials.

One of the major concerns in the development of new ceramic components is a lack of knowledge regarding the nature of the flaws that can ultimately lead to failure. Many of the papers in this volume address this question, as well as the question of the extent to which data obtained on large cracks in fracture mechanics specimens can be used to predict the behavior of "real" flaws. The use of crack growth rate data in lifetime prediction and proof-test schemes is also emphasized.

The field of structural reliability prediction is a fast-moving one. Even as this book goes to print, the methods of data acquisition and analysis are being further refined. Nevertheless, the editors feel that this volume provides a very useful compilation of papers describing the current state of the science in this field.

Stephen W. Freiman

National Bureau of Standards, Washington, D.C. 20234; symposium chairman and editor.

C. Michael Hudson

NASA Langley Research Center, Hampton, Va. 23665; symposium chairman and editor.

David B. Marshall[1]

Failure from Contact-Induced Surface Flaws

REFERENCE: Marshall, D. B., **"Failure from Contact-Induced Surface Flaws,"** *Methods for Assessing the Structural Reliability of Brittle Materials, ASTM STP 844,* S. W. Freiman and C. M. Hudson, Eds., American Society for Testing and Materials, Philadelphia, 1984, pp. 3-21.

ABSTRACT: The scattering of acoustic waves by surface cracks is used in ceramics as both a method of nondestructive evaluation and a means of investigating the mechanics of failure from surface damage. Initially, experiments combining acoustic scattering, *in situ* optical observations, and fracture surface observations of controlled indentation flaws provide essential insight into the scattering process and the mechanics of failure. With more complex flaw configurations, such as machining damage, acoustic scattering measurements provide a unique method for examining the micromechanics of failure and thereby establishing a basis for strength prediction. The results indicate important differences between indentation flaws and ideal stress-free flaws, both in their response to applied loading and in their acoustic scattering characteristics. The differences are due to the influence of residual stresses associated with indentation flaws. Machining-induced cracks behave similarly to indentation cracks. A basis for failure prediction from acoustic scattering measurements can be established for indentation cracks and machining cracks but not for ideal stress-free flaws.

KEY WORDS: failure, strength, machining, scratching, indentation, residual stress, nondestructive testing, acoustic scattering, fractography, structural reliability, brittle materials

Valuable insight into the mechanism of failure from surface flaws in brittle materials has been provided by studies of idealized model flaw systems produced by indentation (for example, Vickers or Knoop). These studies have demonstrated that residual stresses are generated by any mechanical contact damage involving irreversible deformation. The residual stresses dominate the cracking associated with the contact during both crack formation and subsequent loading of the cracks to failure. Consequently, the strength of a dam-

[1]Research engineer, Structural Ceramics Group, Rockwell International Science Center, Thousand Oaks, Calif. 91360.

aged surface is not related exclusively to the size of the largest crack produced by the damage, as in the conventional view of failure; rather the strength is dictated by the residual stresses, which are determined by the contact parameters (load, geometry) and the elastic/plastic response of the material during the contact event. Detailed fracture mechanics analyses for indentation cracking have been developed and verified experimentally by direct observations of flaw response [1-5].

Application of the residual stress concepts derived for isolated indentation flaws to more complex configurations such as machining damage has been demonstrated by observing the scattering of surface acoustic waves from the cracks associated with the damage. In addition to providing a method for identifying the existence of residual stresses and their dominant role in the failure process, the acoustic scattering experiments establish the basis for a method of nondestructive strength prediction.

The main purposes of this paper are to review the current understanding of the mechanisms of failure from contact-induced surface flaws, with particular emphasis on the damage generated by multipoint surface grinding, and to assess the feasibility of nondestructive evaluation using the scattering of acoustic waves. In addition, some new measurements of surface residual stresses associated with machining damage will be presented.

Isolated Cracks

Mechanics of Failure

The importance of residual stresses in the contact-induced cracking of brittle surfaces is readily demonstrated by observing crack evolution during the controlled loading and unloading of well-defined indenters on optically transparent materials. For sharp indenters such as the Vickers or Knoop, the final crack configurations (Fig. 1) are achieved as the indenter is *removed* from the surface [1,3], thus establishing that the driving force for crack formation is provided by a residual stress field. Moreover, since the residual field persists after the contact event, it must supplement any applied loading in driving the cracks to failure. The existence of a postindentation crack-opening force has also been demonstrated by observations of subcritical extension of indentation cracks after indenter removal in materials that are susceptible to environmentally assisted slow crack growth [5,6].

Determination of the stress intensity factor, K_r, due to the residual field is central to any fracture mechanics analysis involving indentation cracks. The residual field results from the elastic/plastic nature of the deformation beneath the indenter and may be evaluated in terms of an outward-acting pressure at the boundary of the plastic zone [1,3]. For approximately axisymmetric indenters, such as the Vickers pyramid, the plastic zone occupies an almost hemispherical volume centered beneath the indentation (Fig. 1, bottom). If

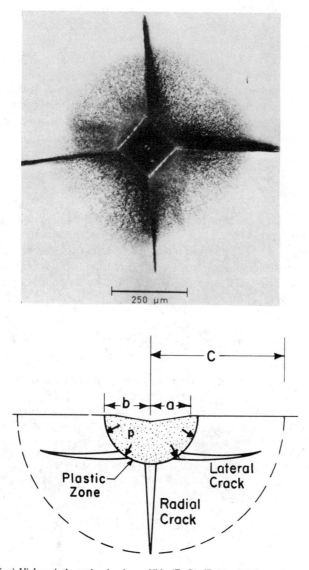

FIG. 1—(Top) *Vickers indentation in zinc sulfide (ZnS)*. (Bottom) *Schematic cross section of the indentation, showing the deformation zone and fractures.*

the crack dimension (c) is sufficiently large compared with the plastic zone radius (b) the pressure may be treated as a point force located at the crack center. Under this condition, a straightforward solution for the stress intensity factor for the radial crack has been derived [1,3]

$$K_r = \frac{\chi_r P}{c^{3/2}} \tag{1}$$

where P is the indenter load and $\chi_r = \S(E/H)^{1/2}$, with E and H the elastic modulus and hardness of the material and $\S$ a dimensionless constant dependent only on indenter geometry. The crack dimension, c_0, after indentation is obtained by equating K_r to the material toughness, K_c, in Eq 1

$$c_0 = \left(\frac{\chi_r P}{K_c} \right)^{2/3} \tag{2}$$

The validity of Eq 2 has been tested with Vickers indentation in a wide range of ceramic materials [5].[2]

The mechanics of failure from radial cracks under the combined influences of the residual stress and a normal applied tension, σ_a, has been analyzed in detail [2,4,7]. The crack response is described by an applied-stress/equilibrium-crack-size function

$$\sigma_a = \left[\frac{K_c}{(\pi \Omega c)^{1/2}} \right] \left[1 - \frac{\chi_r P}{K_c c^{3/2}} \right] \tag{3}$$

(where Ω is a crack geometry parameter), which is obtained by superimposing the stress intensity factors due to the residual and applied fields (K_r from Eq 1 and $K_a = \sigma_a (\pi \Omega c)^{1/2}$) and setting $K_r + K_a = K_c$ for equilibrium crack extension. The failure condition is defined by the maximum in the $\sigma_a(c)$ function

$$c_m = \left(\frac{4\chi_r P}{K_c} \right)^{2/3} = 4^{2/3} c_0 \tag{4}$$

$$\sigma_m = \left[\frac{27}{256} \frac{K_c^4}{\chi_r (\pi \Omega)^{3/2}} \right]^{1/3} P^{-1/3} \tag{5a}$$

$$= \frac{3K_c}{4(\pi \Omega c_m)^{1/2}} \tag{5b}$$

[2]This analysis requires that the crack dimensions be large compared with the scale of any microstructure. For example, in large-grained polycrystalline ceramics the fracture resistance becomes dependent on crack length and orientation, resulting in severe disruption of the ideal crack pattern of Fig. 1 [5].

and failure is preceded by stable equilibrium crack growth from c_0 to c_m. This behavior contrasts with the response of ideal, stress-free cracks, where crack instability is achieved at a critical applied stress level without precursor extension ($\chi_r = 0$, $c = c_0$ in Eq 3).

The indentation fracture analysis has also been extended to the linear deformation fracture configuration [8, 9]. The analysis predicts a similar crack response under applied load, although the region of stable precursor crack growth is more extensive ($c_m/c_0 = 4$) than for axisymmetric penetration ($c_m/c_0 = 2.5$). The linear-damage analysis applies strictly to cracks generated by the penetration of a wedge indenter. However, the observations by Rice and Mecholsky [10], of semielliptical (rather than linear) cracks beneath scratches and machining grooves (see also the section on Machining Damage) suggest that loading during machining may resemble more closely axisymmetric indentation. Such geometrical deviations from linear geometry would be expected to reduce the ratio c_m/c_0.

Observations of Crack Response

Optical Observations—*In situ* measurements of surface traces of indentation cracks during failure testing (Fig. 2a) have confirmed the existence of stable precursor crack extension according to Eq 3 in a wide variety of ceramic materials (glass [2], silicon [11], glass ceramics [12], and silicon nitride [4,13]). Extensive measurements have been obtained in silicon nitride at various contact loads and indenter geometries [4,13]. The data were presented on a universal plot (Fig. 2b) by expressing Eq 3 in terms of normalized variables $S = \sigma_a/\sigma_m$ and $C = c/c_m$, so that the parameters describing indenter geometry and contact load do not appear explicitly

$$S = \left(\frac{4}{3}\right)^{-1/2} C^{-1/2} \left(1 - \left(\frac{1}{4}\right) C^{-3/2}\right) \tag{6}$$

The crack growth curves for two very different indenter geometries (Vickers and Knoop) are coincident, and both are close to the predicted curve,[3] thus illustrating that Eq 3 applies to a wide range of contact configurations.

Acoustic Scattering Observations—The occurrence of stable crack extension prior to failure from contact-induced flaws provides a convenient indication of the existence of residual crack-opening stresses. For indentation cracks, optical observation of radial surface traces, during load application, has confirmed the expected crack response. However, optical observation of cracks in more general damage configurations such as machining is not always possible.

---

[3]The increase of crack length with applied stress becomes rapid as σ approaches σ_m. Confirmation that all of the data in Fig. 2b represent stable equilibrium cracks was obtained by directly observing the cracks while the applied stress was held constant at each measurement point.

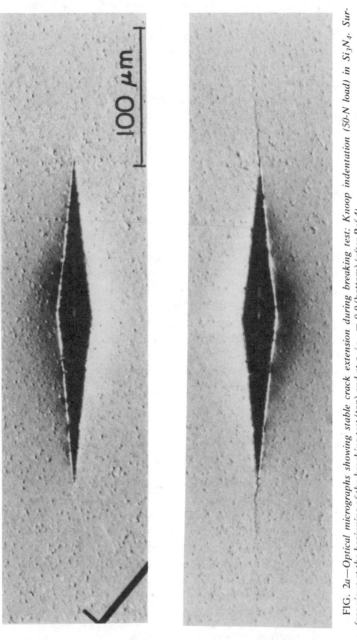

FIG. 2a—Optical micrographs showing stable crack extension during breaking test: Knoop indentation (50-N load) in Si_3N_4. Surface views at the beginning of the breaking test (top) and at $\sigma_a/\sigma_m = 0.9$ (bottom) (after Ref 4).

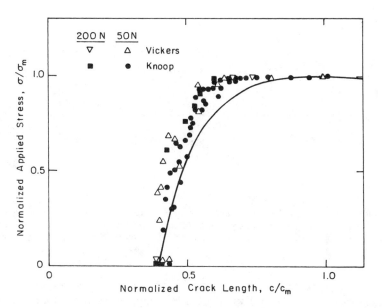

FIG. 2b—*Surface trace measurements of stable crack extension during breaking test; Si₃N₄ bars were indented with Vickers or Knoop indenters and broken in bending (after Ref 4).*

In these cases techniques of crack detection based on the scattering of acoustic waves [14] provide a means of monitoring crack response and thereby determining the influence of residual stresses.

An acoustic scattering technique designed specifically for the detection of surface cracks [15] is illustrated in Fig. 3; transducer 1 excites surface (Rayleigh) waves incident nearly normal to the crack surface, and transducer 2 detects the backscattered waves. The relative amplitude of the backscattered signal is related, by means of scattering analysis, to the crack dimensions, whereas the time delay between the generation and the receiving of the signal defines the crack position.

The acoustic scattering from surface cracks is related uniquely to the crack area, provided the crack surfaces are separated. However, the scattering is sensitive to the existence of crack closure effects. This sensitivity is demonstrated by comparing the acoustic scattering from an indentation crack and an initially stress-free crack[4] of similar dimensions (Fig. 4a). Optical observations confirmed that the stress-free crack did not extend prior to failure. However, the reflected acoustic signal (expressed in Fig. 4a in terms of a calculated crack radius, assuming an *open*, surface half-penny crack [16]) shows a *reversible* increase with applied load. This increase was interpreted in

[4]The stress-free crack was obtained by removing the plastic zone (and therefore the residual stress) of an indentation crack by mechanical polishing. Similar acoustic scattering results have also been obtained from cracks which had the residual stress eliminated by annealing [15].

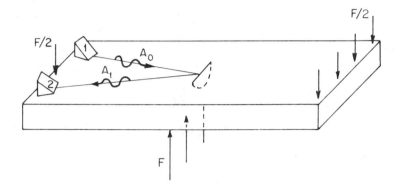

FIG. 3—*The acoustic scattering and mechanical loading configurations used for monitoring crack growth during failure testing:* A_0 = *amplitude of wave excited by transducer 1;* A_1 = *amplitude of scattered wave received by transducer 2;* F = *applied bending force (after Ref 17).*

terms of a reversible opening and closing of the crack surfaces under the applied loading [*15*]. At zero applied stress, complete crack closure is prevented by contacts at asperities over the crack surface. The areas between the contacts scatter as small open cracks of area A_i but, since the scattered amplitude from each open area is approximately proportional to $A_i^{3/2}$ the total scattered amplitude is considerably smaller than that of a fully open crack. Applied tension relieves the contacts continuously until, at the failure point, the crack faces are fully separated and the true crack radius is measured (compare the optical crack length measurement, Fig. 4a).

Acoustic scattering from indentation cracks (which are subject to residual crack opening) does not show the reversible opening and closing effects (Fig. 4b). However, an *irreversible* increase in acoustic signal with applied tension, corresponding to genuine stable crack extension, is detected. Despite some complication in modeling the crack geometry for acoustic scattering analysis,[5] a true measure of the crack dimension is obtained at all stages during the failure test. Comparison of acoustic measurements, optical measurements, and fracture mechanics predictions (Eq 3) are shown in Fig. 4c. The irreversibility of the acoustic scattering response with applied loading provides a definitive indication of the presence of residual crack opening stresses.

The responses of two linear isolated damage configurations (row of indentations, scratch) have also been investigated [*17*]. An irreversible increase in scattered intensity was observed in both cases, thus indicating the existence of stable precursor crack extension due to residual stresses.

---

[5]The crack does not penetrate the plastic zone; therefore, the crack exhibits the geometry of a semiannulus with inner radius dictated by the plastic zone radius. Calculations based on a subsurface elliptical crack have provided a good approximation [*15*].

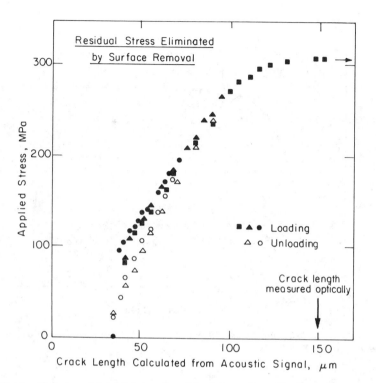

FIG. 4a—*Variation of acoustic scattering, from indentation cracks in polished surfaces of Si₃N₄, during tensile loading: stress-free crack. Note the* reversible *increase in acoustic scattering (expressed as crack length calculated for an open half-penny surface crack) with applied tension.*

Fracture Surface Observations—In some materials the regions of stable and unstable crack extension can be distinguished in optical observations of the fracture surface. The distinction arises from changes in fracture morphology [17] (for example, transgranular to intergranular) or from small perturbations in the plane of propagation [1]. The fracture surface of a Knoop indentation crack in Si₃N₄ is shown in Fig. 5. The reflectivity (brightness) is high in the regions of crack formation and postfailure extension but low in the intermediate region of stable crack growth during loading.

The fracture surface for a row of Knoop indentation cracks in Si₃N₄ is shown in Fig. 6. Under the influence of the applied tension, some of the cracks coalesced and extended stably to an elongated semielliptical surface crack configuration at failure. Similar crack configurations were observed on fracture surfaces resulting from scratch-induced failures [17]. The identification of stable precursor crack growth is consistent with the acoustic scattering results.

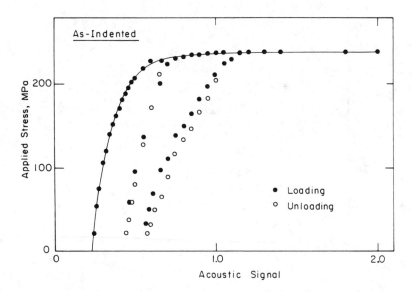

FIG. 4b—*Knoop indentation crack (50-N load). Note the* irreversible *increase in acoustic scattering.*

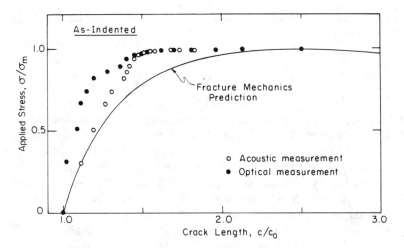

FIG. 4c—*Knoop indentation crack (50-N load): comparison of acoustic measurements, in-situ optical measurements, and fracture mechanics prediction of the variation of crack length with applied tension (after Ref 17).*

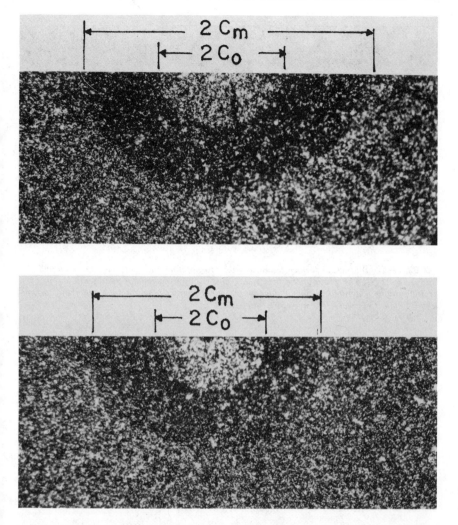

FIG. 5—*Fracture surfaces in Si₃N₄ (width of field 830 μm):* (Top) Knoop indentation (50-N load) in a polished surface (specimen from Fig. 4b). (Bottom) Knoop indentation (50-N load) in a machined surface (after Ref 17).

Machining Damage

Observations of Crack Response

With the acoustic scattering setup of Fig. 3, separate reflected signals were obtained from the cracks associated with the major grooves on machined surfaces of Si_3N_4 [17]. The variation of acoustic scattering from the strength-controlling crack during a failure test is shown in Fig. 7. The irreversible increase in scattered intensity indicates that a residual crack-opening stress

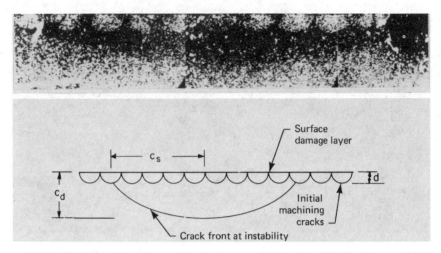

FIG. 6—(Top) *Fracture surface resulting from a row of indentations (50-N load) on a polished surface of Si₃N₄. The width of the field is 2.8 mm.* (Bottom) *Schematic representation of crack configurations generated by linear damage processes (row of indentations, scratching, or machining) and the crack front at failure (after Ref 17).*

caused stable crack growth during loading. This conclusion was supported by fracture surface observations, which showed crack configurations very similar to those in Fig. 6 (due to a row of indentations) with a clearly identifiable row of cracks beneath the grinding groove and a region of stable crack growth. Thus, the response of the strength-controlling cracks in a machined surface appears to follow closely the response of cracks in isolated linear damage configurations. However, the strength of a machined surface is also influenced by the overlap of residual stress fields due to neighboring machining grooves.

Influence of Multiple Grinding Grooves

An isolated grinding groove (or indentation) is surrounded by a plastic zone, which accommodates the volume of the groove (Fig. 1, bottom). The residual stress, which can be evaluated in terms of an outward-acting pressure at the boundary of the plastic zone [2], creates compression adjacent to, and within, the zone and tension on median planes beneath the zone. The cumulative effect of many neighboring damage sites of similar depths, and with a high degree of overlap in their residual fields, would be the development of a uniform thin layer of residual compression (to the depth of the plastic zones) and an underlying residual tension of relatively low magnitude. However, the strength-controlling damage in a machined surface is expected to extend to a greater depth than the average damage in neighboring regions. Thus, the upper portion of the outward-acting pressure from the strength-controlling groove might be negated by a surrounding layer of residual compression from

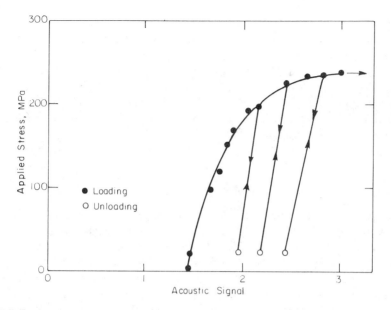

FIG. 7—*Variation of acoustic scattering with applied tension for the strength-controlling flaw in a machined surface of Si₃N₄ (after Ref 17).*

neighboring grooves, but the opening force associated with the lower portion persists [*17*].

The existence of a compressive surface layer in a machined surface was first demonstrated by Cook et al [*18*], by measuring the strengths of glass ceramic flexure bars with indentation cracks introduced into polished and machined surfaces. At identical indentation loads the machined surfaces exhibited higher strengths than the polished surfaces. Similar experiments have been done with scratches and rows of indentations in polished and machined surfaces of Si₃N₄ [*17*]. In all cases the strength of the machined surface was higher than that of the polished surface subjected to the equivalent strength-controlling contact damage. The strength increase was consistently higher for transversely machined bars than for longitudinally machined bars, indicating that the compression is higher in the direction normal to the machining grooves. The strength increase is also sensitive to the size of the strength-controlling flaw in relation to the depth of the machining damage, the largest increase (310 to 530 MPa) being observed for the smallest strength-controlling flaws.

Although a residual compression capable of increasing the strength of a given contact damage by up to 70% has been identified in machined surfaces, it must be emphasized that it is the localized residual tension that exerts a dominating influence on the strength-controlling flaw. This is illustrated in Fig. 5, where the fracture surfaces from Si₃N₄ flexure bars with Knoop indentations in polished (Fig. 5, top) and machined (Fig. 5, bottom) surfaces are compared. In both cases stable crack growth preceded failure (also confirmed by *in-situ* acoustic scattering measurements) [17], indicating that a residual opening stress existed. However, both the initial crack length, c_0, and the extent of stable crack growth, c_m (measured along the surface), are smaller for the machined specimen than for the polished specimen. These observations are consistent with the higher strength measured in the machined specimen (290 MPa compared to 240 MPa).

Measurement of Residual Compression

A quantitative measure of the residual compressive surface layer can be obtained from the degree of elastic bending caused by the layer in a thin plate. The measurement is obtained by first preparing a flat polished surface on one side of a thick plate, and then bonding the polished surface to a rigid support base and reducing the thickness of the plate by machining from the opposite surface. When the plate is removed from the support base, the compression in the machined surface causes the plate to bend so that the polished surface becomes concave (Fig. 8, top). Measurement of the radius of curvature, ρ, by optical interference methods allows the product of the average compression, σ_R, and the thickness, t, of the layer to be evaluated from the relation [19]

$$\sigma_R t = \frac{Ed^2}{6\rho(1 - \nu)} \tag{7}$$

where d is the thickness of the plate ($d \gg t$), E is the elastic modulus, and ν the Poisson's ratio.

An optical interference micrograph of a thin plate of Si_3N_4 prepared in this manner is shown in Fig. 8 (bottom). The elliptical shape of the interference rings indicates that the compression is not equi-biaxial; the compression is maximum (that is, radius of curvature, ρ, is minimum) normal to the machining direction, in agreement with the implication of the strength measurements discussed in the previous section. From Fig. 8 (bottom) we obtain $\rho = 2.2$ m parallel to the machining direction and $\rho = 1.4$ m normal to the machining direction.[6] Then, with $d = 0.340$ mm, $E = 300$ GPa and $\nu = 0.25$, Eq 7 yields $\sigma_R t = 3.5 \times 10^3 \, Pa \cdot m$ parallel to the machining direction and $\sigma_R t = 5.5 \times 10^3 \, Pa \cdot m$ normal to the machining direction.[7]

---

[6]Similar optical interference measurements prior to the machining step indicated that any deviation of the polished surface from perfect flatness was negligible ($< 1 \, \mu$m).

[7]Equation 7 applies to uniform, equi-biaxial compression. However, the corresponding expression for uniaxial compression in a beam differs from Eq 7 by only a factor of $1 - \nu$). Therefore, the error in the present calculations due to the application of Eq 7 to unequal biaxial compressions is expected to be small.

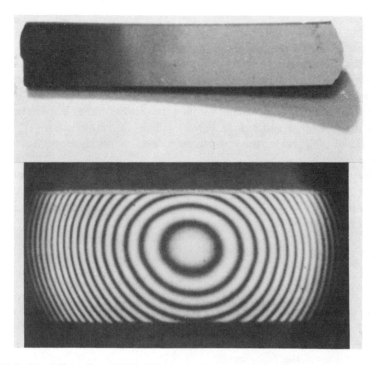

FIG. 8—(Top) *Thin plate of Si₃N₄ (thickness is 0.1 mm) with the upper surface polished (initially flat) and the lower surface subsequently machined.* (Bottom) *Optical interference photograph of the polished surface of a plate similar to that above (thickness is 0.34 mm). The machining direction on the lower surface is horizontal. The wavelength of illumination is 546 nm; the width of field, 10.7 mm.*

Evaluation of σ_R requires a measurement of the thickness of the compressive layer. This could be obtained directly by measuring the change of ρ with removal of the machining damage by polishing, etching, or ion milling. However, in the absence of such measurements, a preliminary estimate of t is obtained here from measurements of plastic zone depths in controlled indentation experiments. For Knoop indentation in Si₃N₄ the plastic zone depth was found to be approximately equal to the width of the residual contact impression [20]. In other experiments [17], a scratch produced by dragging a Knoop indenter across a Si₃N₄ surface, under a normal load of 5 N, left a track of ≈ 10 μm width and degraded the strength by about the same amount as the machining damage. Therefore, if we assume that the ratio of plastic zone depth to contact width is about the same for sliding and stationary Knoop indentation, the depths of the plastic zones associated with the 5 N Knoop scratch and the strength-controlling machining groove are both ≈ 10 μm. Taking this as an upper bound estimate for t, the average compressive stresses become $\sigma_R \geq 350$ MPa parallel to the machining grooves and $\sigma_R \geq 550$ MPa normal to the machining grooves. Notwithstanding the uncertainty in the estimated value of t, these stresses are considerably lower than the compression that exists at the elastic/plastic boundary of an isolated indentation ($\sim H/6$ = 3000 MPa for Si_3N_4—see the Appendix). This result suggests that the compression may have been significantly relieved by material removal during machining.

Residual surface compression of similar magnitude and extent has been detected in machined surfaces of polycrystalline aluminum oxide (Al_2O_3) by Lange et al [21]. The stress was evaluated from X-ray measurements of the change in lattice parameter due to the compressive strain, and the depth of the compressive layer was estimated by taking X-ray measurements after removing various amounts of the machining damage by polishing. By using chromium-radiation with characteristic penetration depth of ~ 8 μm, an average compression σ_R = 170 MPa over a depth ~ 10 μm was found.

Discussion

Implications for Nondestructive Evaluation

The acoustic wave scattering technique was developed primarily as a method of nondestructive evaluation. The results discussed in the previous sections provide essential information for defining the fundamental validity and limitations of the technique.

Two steps are involved in the prediction of strength from ultrasonic measurements of surface cracks. First, the size of the largest crack, c_0, is evaluated from analysis of the acoustic scattering measurements (in the absence of applied loading); then c_0 is related to strength using fracture mechanics. For stress-free cracks, the apparent crack length measured acoustically in the absence of applied loading is not related in a straightforward way to the true crack length (Fig. 4a).[8] A valid measure of the true crack length (which dictates the strength) is obtained only at the point of failure, where the crack surfaces are fully separated by the applied loading. Therefore ultrasonic measurements of stress-free cracks do *not* appear to provide a sound basis for strength prediction. (It is noted, however, that, in the case of Si_3N_4, a conservative strength prediction was obtained by treating a stress-free crack as an indentation crack in both the scattering and the fracture mechanics analyses [15].) For indentations, scratches, and machining damage, on the other hand, the cracks are held fully open by the residual stress[9] in the absence of applied loading. Therefore, provided scattering analysis can be performed for the pertinent crack geometry [15,17,22], the acoustic measurements provide a

---

[8]Analysis of the crack separation process has been performed by Budiansky (1982), but the relation between the true and apparent crack lengths is sensitive to many material parameters (fracture surface topography, grain size, thermal expansion anisotropy, elastic modulus) and the crack size.

[9]As indicated by the absence of significant reversibility in the increase of acoustic scattering with applied loading.

true indication of the crack length and, thus, a fundamentally sound basis for strength prediction.

The fracture mechanics relations required for strength prediction from ultrasonic measurements of indentation cracks in polished surfaces are given in Eqs 4 and 5b (the initial crack length, c_0, is related to the crack length c_m at the failure point by Eq 4, and c_m is related to the strength by Eq 5b). Strength prediction for machining damage and scratches requires analagous relations. However, the deformation/fracture geometry of Fig. 6 (bottom) is not amenable to straightforward analysis. Consequently, a semiempirical approach has been employed to derive the requisite relations [17]. Measurements of crack dimensions from fracture surfaces in Si_3N_4 indicated that the extent of prefailure crack extension was approximately constant for machining damage, scratches, and rows of indentations at various strength levels.

$$\frac{c_m}{c_0} \approx 5 \tag{8}$$

where $c_m = (c_s c_d)^{1/2}$ (Fig. 6) is the characteristic crack dimension at the failure point. Moreover, the strengths σ_m for the same set of specimens were related to c_m by[10]

$$\sigma c_m^{1/2} = 3.9 \, \text{MPa} \cdot \text{m}^{1/2} \tag{9}$$

The application of Eqs 8 and 9 to predict strengths of Si_3N_4 from acoustic measurements, obtained both with the experimental setup described in this paper and with another setup that was designed to permit scanning of the entire specimen surface, is described elsewhere [15,17,22].

Implications for Damage Resistance

The competing influences of the strength-degrading dominant flaw and the compressive surface damage layer in a machined surface present a possibility to optimize the machining procedure for a given application. Generally, the strength of a machined surface would be expected to decrease with increasing severity of machining (large abrasive particles, high machining forces), but the depth of the compressive layer would be expected to increase. The compressive layer provides resistance to in-service strength degradation from mechanical contact events. Therefore, for structural applications in mechanically hostile environments, optimum performance could be provided by the most severe machining procedure (giving maximum resistance to in-service mechanical

[10]It is noted that the similarity between Eqs 8 and 9 and the corresponding relations for indentation cracks (Eqs 4 and 5b) might be expected on the basis that the replacement of K_r in Eq 1 with any function of the form $K_r = \chi_r P/c^n (n > 0)$ yields a set of equations of the same form as Eqs 2 to 5 but with numerical factors dependent upon n.

damage) that maintains the strength of the machining damage above some minimum requirement.

APPENDIX

The Residual Pressure at the Elastic/Plastic Boundary in Vickers Indentation

A measure of the residual pressure, p, acting at the elastic/plastic boundary of an isolated Vickers indentation (Fig. 1) can be obtained from measurements of the extent of cracking caused by the residual stress field. In the analysis described in the section on Isolated Cracks, the expression that led to Eq 1 was [1,4]

$$K_r = \frac{2P_r}{(\pi c)^{3/2}} \tag{10}$$

where P_r is the residual wedging force, due to the pressure p, located at the crack center. With $P_r = \pi b^2 p/2$, $K_r = K_c$, and the hardness relation $H = P/2a^2$ (where P is the indenter load and a the half diagonal of the indentation), Eq 10 can be written

$$p = \frac{\pi^{1/2} K_c c^{3/2}}{b^2} = 2\pi^{1/2} \left(\frac{a}{b}\right)^2 K_c \left(\frac{c^{3/2}}{P}\right) H \tag{11}$$

Then, with the following previously published data for Si_3N_4, $b/a \approx 1.2$ [23], $K_c = 4$ MPa $\cdot$ m$^{1/2}$ [5], $P/c^{3/2} = 56$ MPa $\cdot$ m$^{1/2}$, and $H = 18$ GPa[4], the residual pressure becomes $p \approx H/6 = 3000$ MPa. This pressure agrees well with the value calculated from a model based on an internally pressurized spherical cavity [23]

$$p \approx H \left\{ \left[1 + 3 \ell n \left(\frac{b}{0.45a} \right) \right]^{-1} - \left(\frac{0.45a}{b} \right)^3 \right\} \approx \frac{H}{5} \tag{12}$$

References

[1] Marshall, D. B. and Lawn, B. R., *Journal of Materials Science*, Vol. 14, No. 8, 1979, pp. 2001–2012.

[2] Marshall, D. B. and Lawn, B. R., *Journal of Materials Science*, Vol. 14, No. 9, 1979, pp. 2225–2235.

[3] Lawn, B. R., Evans, A. G., and Marshall, D. B., *Journal of the American Ceramic Society*, Vol. 63, Nos. 9–10, 1980, pp. 574–581.

[4] Marshall, D. B., *Journal of the American Ceramic Society*, Vol. 66, No. 2, 1983, pp. 127–131.

[5] Anstis, G. R., Chantikul, P., Lawn, B. R., and Marshall, D. B., *Journal of the American Ceramic Society*, Vol. 64, No. 9, 1981, pp. 533–538.

[6] Gupta, P. K. and Jubb, N. J., *Journal of the American Ceramic Society*, Vol. 64, No. 8, 1981, pp. C112–C114.

[7] Chantikul, P., Anstis, G. R., Lawn, B. R., and Marshall, D. B., *Journal of the American Ceramic Society*, Vol. 64, No. 9, 1981, pp. 539–543.

[8] Kirchner, H. P. and Isaacson, E. D., in *Fracture Mechanics of Ceramics*, Vol. 4, R. C. Bradt, D. P. H. Hasselman, F. F. Lange, and A. G. Evans, Eds., Plenum, New York, 1983, p. 57.

[9] Kirchner, H. P. and Isaacson, E. D., *Journal of the American Ceramic Society*, Vol. 65, No. 1, 1982, pp. 55-60.

[10] Rice, R. W. and Mecholsky, J. J., in *The Science of Ceramic Machining and Surface Finishing II*, Special Publication, No. 562, B. J. Hockey and R. W. Rice, Eds., National Bureau of Standards (U.S.), Washington, D.C., 1979, pp. 351-378.

[11] Lawn, B. R., Marshall, D. B., and Chantikul, P., *Journal of Materials Science*, Vol. 16, No. 7, 1981, pp. 1769-1775.

[12] Cook, R. F., Lawn, B. R., and Anstis, G. R., *Journal of Materials Science*, Vol. 17, No. 4, 1982, pp. 1108-1116.

[13] Marshall, D. B., in *Progress in Nitrogen Ceramics*, F. L. Riley, Ed., Nijhoff, The Hague, The Netherlands, 1983.

[14] Khuri-Yakub, B. T., Kino, G. S., and Evans, A. G., *Journal of the American Ceramic Society*, Vol. 63, No. 1, 1980, pp. 65-71.

[15] Tien, J. J. W., Khuri-Yakub, B. T., Kino, G. S., Evans, A. G., and Marshall, D. B., *Journal of Non Destructive Evaluation*, Vol. 2, Nos. 3-4, 1981, pp. 219-229.

[16] Kino, G. S., *Journal of Applied Physics*, Vol. 49, No. 6, 1978, pp. 3190-3199.

[17] Marshall, D. B., Evans, A. G., Tien, J. J. W., and Kino, G. S., *Proceedings of the Royal Society of London*, Vol. A385, 1983, pp. 461-475.

[18] Cook, R. F., Lawn, B. R., Dabbs. T. P., and Chantikul, P., *Journal of the American Ceramic Society*, Vol. 64, No. 9, 1981, pp. C121-C122.

[19] Oel, H. J. and Frechette, V. D., *Journal of the American Ceramic Society*, Vol. 50, No. 10, 1967, pp. 542-549.

[20] Mendiratta, M. G. and Petrovic, J. J., *Journal of Materials Science*, Vol. 11, No. 5, 1976, pp. 973-976.

[21] Lange, F. F., James, M. R., and Green, D. J., "Determination of Residual Stresses Caused by Grinding in Polycrystalline Al_2O_3," *Journal of the American Ceramic Society*, Vol. 66, No. 2, 1983, pp. C16-C17.

[22] Khuri-Yakub, B. T., Kino, G. S., Liang, K., Tien, J., Chou, C. H., Evans, A. G., and Marshall, D. B., in *Review Progress in Quantitative Non-Destructive Evaluation*, Vol. 1, D. Thompson and D. E. Chimenti, Eds., Plenum, New York, 1982.

[23] Chiang, S. S., Marshall, D. B., and Evans, A. G., *Journal of Applied Physics*, Vol. 53, No. 1, 1982, pp. 298-311.

Robert F. Cook[1] and Brian R. Lawn[2]

Controlled Indentation Flaws for Construction of Toughness and Fatigue Master Maps

REFERENCE: Cook, R. F., Lawn, B. R., **"Controlled Indentation Flaws for Construction of Toughness and Fatigue Master Maps,"** *Methods for Assessing the Structural Reliability of Brittle Materials, ASTM STP 844,* S. W. Freiman and C. M. Hudson, Eds., American Society for Testing and Materials, Philadelphia, 1984, pp. 22-42.

ABSTRACT: A simple and economical procedure for accurate determinations of toughness and lifetime parameters is described. Indentation flaws are introduced into strength test pieces, which are then taken to failure under specified stressing and environmental conditions. By controlling the size of the critical flaw, by means of the contact load, material characteristics can be represented universally on "master maps" without the need for statistical considerations.

This paper surveys both the theoretical background and the experimental methodology associated with the proposed scheme. The theory is developed for "point" flaws for dynamic and static fatigue, explicitly incorporating load into the analysis. A vital element of the fracture mechanics is the role played by residual contact stresses in driving the cracks to failure. Experimental data on a range of Vickers-indented glasses and ceramics are included to illustrate the power of the method as a means of graphic materials evaluation. It is demonstrated that basic fracture mechanics parameters can be measured directly from the slopes, intercepts, and plateaus on the master maps and that these parameters are consistent, within experimental error, with macroscopic crack growth laws.

KEY WORDS: fatigue, indentation flaw, lifetime prediction, master maps, materials evaluation, strength, toughness, universal curves, structural reliability, brittle materials

The increasing use of glasses and ceramics as structural materials has prompted the development of new and accurate techniques for evaluating intrinsic fracture parameters. Chief among these parameters are the fracture toughness, K_c, and the crack velocity exponent, n, which respectively characterize the equilibrium and kinetic crack growth responses. In the context of

[1] Graduate student, Department of Applied Physics, School of Physics, University of New South Wales, Kensington, N.S.W. 2033, Australia.
[2] Physicist, Center for Materials Science, National Bureau of Standards, Washington, D.C. 20234.

brittle design it is essential to achieve an adequate level of precision in such parameter evaluations. This is particularly so in consideration of component integrity under sustained stresses and chemical environments, where apparently minor uncertainties can translate into order-of-magnitude discrepancies in lifetime predictions.

A standard method of determining basic fracture parameters for design is to measure the strengths of representative test specimens in flexure. However, for specimens with typically as-received or as-prepared surfaces these strengths depend not only on intrinsic material properties but on flaw distributions as well. Under such conditions it is not possible to investigate these two elements of the problem in *any truly independent way*. Evaluation of material parameters becomes a mere exercise in statistical data manipulation, with little or no physical insight into the nature of the critical flaws responsible for failure [1-2]. This probabilistic approach makes it difficult to assess the relative merits of different materials from the standpoint of intrinsic properties alone.

A controlled-flaw technique that effectively eliminates the statistical component from strength testing has been developed in a recent series of articles [3-12]. A single dominant flaw of predetermined size and geometry is introduced into the prospective tensile surface of each specimen using a standard diamond indenter. The specimens are then stressed to failure in the usual way. With the indentation and flexure testing conditions held fixed, any variation in the strength behavior can be taken as a direct reflection of the intrinsic material response. The only need for statistical treatments, then, resides in the trivial accountability of random scatter in the data. Quite apart from the ensuing improvements in data reproducibility, the indentation procedure confers several advantages in strength analysis: (1) greater specimen economy is achieved; (2) the location of the critical flaw is predetermined, thereby allowing for closer observation of the fracture mechanics to failure; (3) indentations provide a reasonable simulation of the damage processes responsible for a great many brittle failures [13-15]. One apparent complication which attends the technique is the existence of a strong residual contact field about the elastic/plastic deformation zone, necessitating the incorporation of additional terms in the governing stress intensity factor. However, closed-form solutions of the fracture mechanics formulations are now available for both equilibrium [4] and kinetic [16] conditions of failure; analytical determinations of toughness and fatigue parameters from the strength data may accordingly be made in as straightforward a manner as for Griffith flaws without the residual stress term.

The capacity to control the scale of the critical flaw through the indentation load is a potent tool in the investigation of material fracture properties. The load actually replaces initial crack size as a variable in the fracture equations, thereby eliminating the need for onerous measurements of crack dimensions (although some observations of crack growth are useful for confirming the validity of the theory) [15]. Size effects in the micromechanics may then be studied systematically: important changes in the nature of low-load contact flaws have been thus revealed on reducing the crack size to the scale of the deformation zone [17] or the microstructure.[3] Systematic variations in the load dependence of indentation-strength characteristics can also be used to evaluate preexisting stress states in brittle materials, such as tempered glass [18]. Again, some materials may produce ill-defined indentation patterns outside certain ranges of flaw size or be restricted in specimen dimensions, in which case the geometrical requirements of standard strength-testing procedures may make it impossible to operate at a single contact load. The theoretical analysis allows one to compensate for any such changes in the working contact conditions, effectively reducing all data to an "equivalent" load.

This paper illustrates a procedure for representing the intrinsic strength properties of brittle materials on an indentation master map. A suitable normalization scheme incorporating indentation load into the plotting coordinates allows for the reduction of all inert and fatigue strength data onto universal curves for the various test materials. In this sense, the scheme is reminiscent of that developed earlier by Mould and Southwick [19], except that their use of relatively ill-defined abrasion flaws necessitated a totally empirical approach in the data reduction. On our master map, the position of a given curve may be taken as a graphic indicator of the intrinsic toughness and fatigue susceptibility. Quantitative determinations may accordingly be made of K_c and n without recourse to statistically based theories of strength.

Background Theory

Stress Intensity Factor for Indentation Cracks

The starting point in the analysis is the stress intensity factor for an indentation crack of characteristic dimension c produced at peak contact load P and subjected to subsequent applied tensile stress σ_a. For "point" flaws produced in axially loaded indenters, the general form of this stress intensity factor is [4]

$$K = \frac{\chi P}{c^{3/2}} + \psi \sigma_a c^{1/2} \tag{1}$$

where χ and ψ are dimensionless parameters. The second term in Eq 1 is the familiar contribution from the applied field; ψ depends only on crack geometry, here assumed to be essentially pennylike [20]. The first term is the contribution from the residual contact field; for materials which deform irreversibly by a constant volume process

---

[3] R. F. Cook, University of New South Wales, unpublished work, 1983.

$$\chi = \xi \left(\frac{E}{H}\right)^{1/2} \qquad (2)$$

approximately [21], where E is Young's modulus, H is hardness, and ξ is a numerical constant.

In the event of any preexistent stress acting on the crack, a third term would have to be included in Eq 1 [4,9]. Except to note that this potential complication should be heeded when preparing the surfaces of test specimens, we shall consider it no further in our mathematical derivations.

Equilibrium Solutions: Inert Strengths

Equilibrium conditions of crack growth are closely realized experimentally by testing in an inert environment. In terms of fracture mechanics notation, the criterion for equilibrium is that $K = K_c$. If $dK/dc < 0$ the equilibrium is stable; if $dK/dc > 0$ it is unstable. Now it is evident from Eq 1 that K for given values of P and σ_a passes through a minimum in its functional dependence on c; thus at subcritical configurations $K(\min) < K_c$, there is a stable and an unstable equilibrium, to the left and to the right of the minimum, respectively [16]. In an inert strength test, σ_a is increased steadily until these two equilibria merge at $dK/dc = 0$, which defines the critical variables

$$\sigma_m = \frac{3K_c}{4\psi c_m^{1/2}} \qquad (3a)$$

$$c_m = \left(\frac{4\chi P}{K_c}\right)^{2/3} \qquad (3b)$$

at which crack growth proceeds without limit. We may note that any relaxation of the residual stress field, as reflected in a reduction in χ (or, more specifically, in ξ in Eq 2), will cause σ_m to expand and c_m thence to contract.

It can be shown that the ideal indentation crack is in a state of equilibrium immediately after completion of the contact cycle [21]. The size of this crack is found by setting $\sigma_a = 0$, $K = K_c$ in Eq 1

$$c_0 = \left(\frac{\chi P}{K_c}\right)^{2/3} \qquad (4)$$

From Eq 3b we have $c_0 \simeq 0.40 c_m$. On subsequently applying the tensile stress the crack extends stably from c_0 to c_m, whence spontaneous failure ensues at $\sigma_a = \sigma_m$ [4]. In reality, deviations from this ideal behavior are observed; relaxation effects can cause c_m to contract, as already mentioned, and subcritical, moisture-assisted crack extension within the residual contact field can cause c_0 to expand, to c_0', say. Nevertheless, unless the condition $c_0' \leq c_m$ is violated, some precursor crack growth will still precede failure, in which case σ_m remains a measure of the inert strength.

Equation 3 may then be conveniently rearranged to eliminate all terms in crack size, and then combined with Eq 2 to yield

$$\sigma_m P^{1/3} = \left(\frac{3}{4\psi}\right)\left(\frac{1}{4\xi}\right)^{1/3}\left[\left(\frac{H}{E}\right)^{1/8} K_c\right]^{4/3} \tag{5}$$

This expression conveniently relates the test variables on the left side to the material properties, primarily the toughness, on the right side. We emphasize once more that this formulation is contingent on the absence of all spurious prepresent stresses.

Kinetic Solutions: Dynamic Fatigue

When cracks are exposed to moisture or other interactive environmental species, extension can occur in the subcritical region, $K < K_c$. The major characteristic of this kind of extension is its rate dependence, which, in turn, is highly sensitive to the crack driving force. The basic equation of kinetic fracture accordingly takes the form of a crack velocity, $v(K)$. In the interest of obtaining closed-form solutions to the ensuing fracture mechanics relations, we choose the empirical power law function [22]

$$v = v_0\left(\frac{K}{K_c}\right)^n \tag{6}$$

where v_0 and n are material/environment parameters. Materials with lower values of n are said to be more susceptible to kinetic crack growth effects.

The most practical loading arrangement for the systematic study of rate effects in strength properties is that of dynamic fatigue, in which the time differential of stress is held fixed up to the point of failure, that is, $\dot\sigma_a = \sigma_a/t = $ constant. We may thus combine Eqs 1 and 6 to obtain a differential equation for this stressing configuration

$$\frac{dc}{dt} = v_0\left[\frac{\chi P}{K_c c^{3/2}} + \frac{\psi\dot\sigma_a c^{1/2} t}{K_c}\right]^n \tag{7}$$

This equation has to be solved at given values of P and $\dot\sigma_a$ for the time to take the crack from its initial configuration, $K = K(c_0')$, to its final configuration, $K = K_c$, at which point the stress level defines the dynamic fatigue strength, $\sigma_a = \sigma_f$ [16].

$$\sigma_f = (\lambda'\dot\sigma_a)^{1/(n'+1)} \tag{8}$$

where

$$n' = \frac{3n}{4} + \frac{1}{2} \tag{9a}$$

$$\lambda' = (2\pi n')^{1/2} \frac{\sigma_m^{n'} c_m}{v_0} \tag{9b}$$

The solution in Eq 8 is identical in form to that for Griffith flaws ($\chi = 0$) [22]. However, the slopes and intercepts from a linear plot of log σ_f against log $\dot{\sigma}_a$ are very different in the two instances. In the present case ($\chi \neq 0$) n' and λ' may be regarded as apparent fatigue parameters, in the sense that transformation equations are required to convert these to true crack velocity exponent and coefficient terms. Thus, Eq 9a may be inverted to obtain n directly from n', and Eq 9b similarly (in conjunction with measured values of σ_m and c_m) to obtain v_0 from λ'. It is again seen that initial crack size does not enter the results, as long as the condition $c_0' \leq c_m$ remains operative [9].

Implicit in the derivation of Eq 8 is the usual assumption that the prospective test surfaces are free of spurious stresses. The introduction of such stresses leads to nonlinearities in the dynamic fatigue plotting scheme, thereby destroying the basis for the above analysis [9,10].

It is convenient at this point to incorporate the indentation load as a working test variable into the dynamic fatigue relations. Whereas n' in Eq 9a is independent of all test variables, λ' in Eq 9b can be expressed as an explicit function of P through the quantities σ_m and c_m in Eq 3. In this way, we may write

$$\lambda' = \frac{\lambda_P'}{P^{(n'-2)/3}} \tag{10}$$

where λ_P' is a *modified* intercept term, totally independent of P, given by

$$\lambda_P' = \frac{(2\pi n')^{1/2} \left(\dfrac{3K_c}{4\psi}\right)^{n'} \left(\dfrac{K_c}{4\chi}\right)^{(n'-2)/3}}{v_0} \tag{11}$$

Equation 10 tells us that fatigue data obtained on one material but using different indentation loads will fall on different straight lines, mutually translated but without change of slope. Now by inserting Eq 10 into Eq 8 we may appropriately modify the dynamic fatigue relation, thus

$$\sigma_f P^{1/3} = (\lambda_P' \dot{\sigma}_a P)^{1/(n'+1)} \tag{12}$$

so that by plotting log $(\sigma_f P^{1/3})$ against log $(\dot{\sigma}_a P)$ all data should fall onto a universal fatigue curve. This plot would, of course, cut off at a limiting level on the ordinate corresponding to the inert strength plateau defined in Eq 5. The procedure for evaluating crack velocity parameters from the slopes and intercepts of such representations is the same as before, but with Eq 10 serving as an intermediary to Eq 9.

Kinetic Solutions: Static Fatigue

Of more practical interest from a design standpoint is the issue of component lifetime under fixed stress rather than stress *rate*. Ideally, it would seem desirable to formulate a universal static fatigue relation in direct analogy to Eq 12 retaining, as far as possible, the same adjustable parameters. Lifetime predictions could then be made from dynamic fatigue data alone, without having to resort to delayed failure experiments. This formulation may be achieved in two steps. First, eliminate stressing rate in favor of time to failure, $\dot{\sigma}_a = \sigma_f/t_f$. This step introduces the lifetime concept without yet altering the status of Eq 12 as a dynamic fatigue relation. Then, convert to equivalent static fatigue variables by replacing σ_f with σ_A, that is, the level of the invariant applied stress, and t_f with $(n' + 1)t_f$ [16]. The resulting static fatigue relation is

$$\frac{t_f}{P^{2/3}} = \frac{\lambda_P'}{(n' + 1)(\sigma_A P^{1/3})^{n'}} \tag{13}$$

We reiterate here, at the risk of laboring the point, that the variables P, σ_A, and t_f in Eq 13 relate to prospective static fatigue conditions, whereas the parameters n' and λ_P' are adjustables, as defined by Eqs 9 and 10, to be determined from dynamic fatigue data.

Experimental

Materials Selection and Preparation

The materials in this study were chosen in accordance with two major criteria: first, they should cover a range of toughness and crack velocity characteristics, as determined by independent fracture techniques; second, they should be of some technical importance. Table 1 [11,23-27] lists these materials and their pertinent properties.

All the specimens were prepared in the usual manner for strength testing. However, particular attention was paid to surface preparation, bearing in mind our repeated assertion that preexisting stress states can greatly influence the interpretation of strength data. The glass specimens were therefore annealed [18] and the ceramics surface polished to a mirror finish with diamond paste [10] to ensure removal of any such stresses.

TABLE 1—*Materials used in this study.*

Material	Independent Parameters				Indentation Parameters		
	$E,$ GPa	$H,$ GPa	$K_c,$ MPa·m$^{1/2}$	n	$K_c,$ MPa·m$^{1/2}$	n	log $\nu_0,$ ms^{-1}
Soda lime glass[a]	70	6.6	0.74*	16–19*	0.97	18	−1.6
Borosilicate glass[b]	89	6.5	0.77*	31–37*	1.2	36	1.6
Fused silica[c]	72	7.6	0.81*	38*	1.2	44	2.2
Synroc[d]	190	10.3	1.9	...	1.8	35	0.2
PZT[e]	88	3.1	0.87	...	1.0	43	−0.5
Alumina[f]	400	16	4.4	46*	3.8	59	1.7
Silicon carbide[g]	435	24	4.1*	118*	3.7	222	8.4
Glass ceramic[h]	108	8.4	2.5*	63,* 84*	2.2	117	5.0

*Determinations by other workers (see References, below).
[a] Schott-Ruhrglas GMBH [11,23] (S. M. Wiederhorn, National Bureau of Standards, unpublished work, 1983).
[b] Schott-Ruhrglas GMBH [11,23].
[c] Schott-Ruhrglas GMBH [23,24].
[d] Synroc B, Australian Atomic Energy Research Establishment [24].
[e] Lead zircon titanate, Plessey, Australia.
[f] F99, Friedrichsfeld GMBH [25] (A. C. Gonzalez and S. W. Freiman, National Bureau of Standards, unpublished work, 1982).
[g] NC203, Norton Co. [7,26].
[h] Pyroceram C9606, Corning Glass Co. [7,10,27] (B. G. Koepke, Honeywell, unpublished work, 1980).

Indentation and Strength Testing Procedure

All the specimens were routinely indented centrally along their length using a Vickers diamond pyramid indenter to produce dominant flaws for the subsequent failure tests. The Vickers geometry was chosen both for its proven capacity to produce well-defined radial crack patterns and for its general availability in hardness testing facilities. The glasses were indented at several loads, ranging from 0.05 to 100 N, whereas the ceramics were each indented at single loads of 10, 20, or 100 N. In all cases the radial cracks extended well beyond the central hardness impression, but never to a length in excess of one-tenth the specimen thickness.

The indented specimens were then broken in four-point flexure {ASTM Flexure Testing of Glass [C 158-72 (1978)]} in a universal testing machine at constant crosshead speed. Care was taken to center the indentation on the tension side, with one set of radial cracks aligned normal to the long axis. The breaking loads were recorded using conventional strain gage and piezoelectric load cells [10], and the corresponding rupture stresses thence evaluated from simple beam theory. Inert strengths, σ_m, were measured in dry nitrogen or argon or silicone oil environments, with the crosshead running at its maximum speed. Dynamic fatigue strengths, σ_f, were measured in distilled water over the allowable range of crosshead speeds. At least six specimens were broken in each strength evaluation, from which means and standard deviations were computed.

Measurement of Critical Crack Dimensions

For the purpose of confirming the necessary condition that the initial crack size c_0' should never exceed the instability value c_m for equilibrium failure, and for verifying certain aspects of the fatigue solutions presented earlier, an optical examination of representative critical indentations is recommended. The technique used here was to place three indentations instead of one on a given test surface and then take the specimen to failure under inert conditions [10]. On the understanding that all three indentations must have had nearly identical growth histories, the procedure leaves two "dummies" in the broken test piece from which to measure the required crack dimensions. The Vickers geometry proves particularly useful in this technique, for while the set of radial cracks perpendicular to the tensile direction provides a measure of c_m, the set parallel to this same direction remains free of external stress and hence provides a measure of c_0'.

In all the materials studied in this work, some precursor crack growth was indeed found to occur prior to failure.

Results

Inert Strengths and Toughness

In this section we begin by examining the dependence of inert strength on indentation load for the three glasses studied. With this dependence established, we then investigate how the inert strength data may be reduced to a composite toughness parameter for all of the test materials.

Figure 1 accordingly shows σ_m as a function of P for the glasses. The straight lines are best fits of slope $-\frac{1}{3}$ in logarithmic coordinates, as in Eq 5. This same dependence has been confirmed elsewhere for several other brittle materials [7,28,29].[3]

Values of the composite parameter $\sigma_m P^{1/3}$ are thus evaluated for each of the glasses and ceramics and are plotted as a function of $(H/E)^{1/8}K_c$ (from Table 1) in Fig. 2. The straight line is a fit of logarithmic slope $\frac{4}{3}$ in accordance with Eq 5, using a calibration value $(3/4\psi)(1/4\xi)^{1/3} = 2.02$ from an earlier, more comprehensive study [7]. The trends in Fig. 2 appear to be in reasonable accord with prediction, although some deviations are evident, particularly for the fused silica and borosilicate glasses. Estimates of the "indentation toughness" obtained directly from $\sigma_m P^{1/3}$ by inverting Eq 5 are included in Table 1 for comparison with the independently determined values.

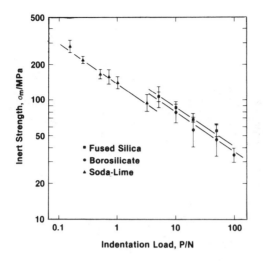

FIG. 1—*Inert strength as function of indentation load for the silicate glasses. (Data courtesy T. P. Dabbs.)*

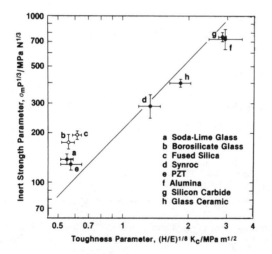

FIG. 2—*Inert strength parameters, $\sigma_m P^{1/3}$, as a function of the toughness parameter, $(E/H)^{1/8} K_c$, for the glasses and ceramics.*

Dynamic Fatigue and Crack Velocity Parameters

We consider now the dynamic fatigue responses, again beginning with the glasses to examine the functional influence of contact load, and outline the procedure for determining the exponent and coefficient in the crack velocity function.

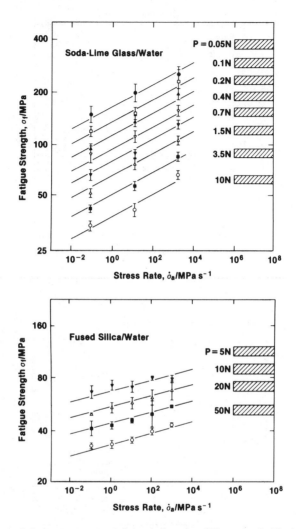

FIG. 3—*Dynamic fatigue responses of glasses indented at different loads. The hatched bands indicate inert strength levels. (Data courtesy T. P. Dabbs.)*

Figure 3 shows these responses for the glass compositions in water. The straight lines drawn through individual sets of data at fixed P are best fits to Eq 8, regressed for each glass on *all* the data consistent with the intercept relation Eq 10. Thus we obtain families of lines of constant slope, with systematic displacements to lower strength levels with increasing load. Analogous plots are shown in Fig. 4 for the five ceramics in the same water environment, but now for a single load in each case. The inert strength limits are included in all plots as a reference baseline for assessing the degrees of fatigue.

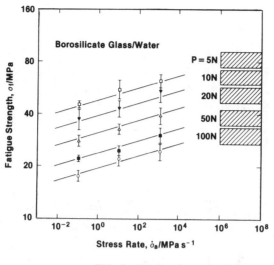

FIG. 3—*Continued.*

From the regressed slopes and intercepts we obtain values of the apparent fatigue parameters n' and λ' in Eq 8. Inversion of Eq 9 (together with the inert strength data) then allows us to evaluate the true crack velocity parameters, n and v_0. These evaluations are summarized in Table 1; comparisons may be made in this tabulation with independent measurements of the crack velocity exponent.

Master Maps

We have set the base for determining universal fracture curves for the materials studied, and thence to construct master maps. We do this for dynamic and static fatigue conditions in turn.

The presentation of the dynamic fatigue results on a single master map requires conversion of all data to appropriate load-adjusted variables $\sigma_f P^{1/3}$ and $\dot{\sigma}_a P$ in Eq 12. Figure 5, an appropriate composite of all data thus converted from Figs. 3 and 4 (but with the error bars omitted for clarity), is such a map. Each material is now conveniently represented by a universal curve, independent of the contact loads used to obtain the data. The curves plotted in this diagram represent numerical solutions of the basic fatigue differential equation (Eq 7), obtained for the ranges of P and $\dot{\sigma}_a$ covered experimentally for each material, using the inert and kinetic parameters already determined along with the measured initial crack sizes [10]. The fact that the curves regenerated in this way are effectively coincident with the data is, of course, no real surprise, since the regression analyses used in the parameter evaluations were performed in accordance with the solutions of the differential equation

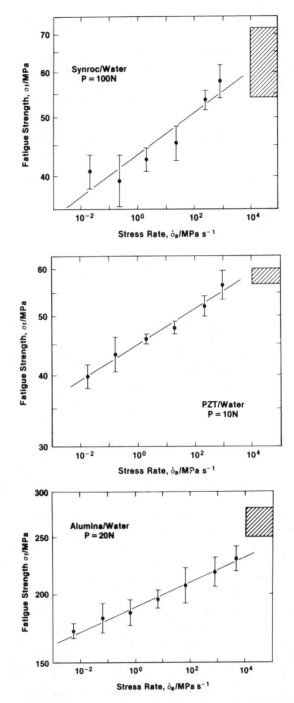

FIG. 4—*Dynamic fatigue responses of ceramics indented at single loads. The hatched bands indicate inert strength levels.*

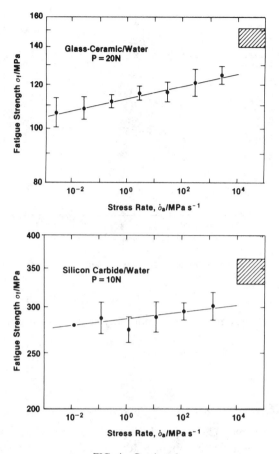

FIG. 4—*Continued.*

in the first place. An exercise of this kind nevertheless serves two useful purposes: (1) to confirm that the solutions referred to, which are of closed form, are indeed reasonably reliable; (2) to show how closely the curves remain linear in the fatigue region and then plateau out at the inert strength levels $\sigma_m P^{1/3}$ (Fig. 2).

The equivalent construction for static fatigue is obtained from the constant stressing rate results using the rationale described earlier in the derivation of Eq 13. Thus we generate the plots shown in Fig. 6 directly from the best-fit values of n' and λ' (or more strictly, in Eq 10, λ_p') determined by the data regressions shown in Figs. 3 and 4. Cutoff levels on the abscissa again correspond to inert strength limits. Because the construction in Fig. 6 is not obtained this time from regenerated solutions of the basic differential equation, we are unable to plot the curved transition between the fatigue and inert regions; however, the abruptness of the corresponding crossover points in Fig. 5

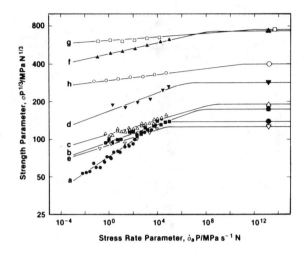

FIG. 5—*Dynamic fatigue master map. Error bars are omitted from the data points for clarity. See Table 1 for a key to the materials.*

suggests that we may reasonably ignore any such curvature in the lifetime maps.

Discussion

Quantitative Evaluation of Fracture Parameters

The scheme presented here for reducing fatigue data to universal curves for any specified material/environment system, and thence for constructing master maps to facilitate comparisons between these curves, provides an attractive route to simple, accurate, and economical evaluation of fracture parameters for design. In the following subsections we shall discuss how these constructions may be used as a quantitative tool for parameter determinations in different regions of the curves.

Inert Strength Levels—The position of the inert strength cutoff level, $\sigma_m P^{1/3}$, may be taken as an indicator of material toughness, K_c. Intrinsically tougher materials will therefore exhibit cutoffs further toward the top of a dynamic fatigue map (Fig. 5) and toward the right of a static fatigue map (Fig. 6).

It should be emphasized that the correspondence implied here is not exact. To clarify this point we may invert Eq 5 to obtain an explicit expression for toughness

$$K_c = \left(\frac{256\psi^3\xi}{27}\right)^{1/4}\left(\frac{E}{H}\right)^{1/8}(\sigma_m P^{1/3})^{3/4} \qquad (14)$$

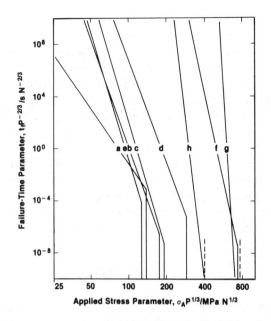

FIG. 6—*Static fatigue master map, generated from Fig. 5. See Table 1 for a key to the materials.*

Thus, K_c depends on the elastic/plastic term E/H as well as on $\sigma_m P^{1/3}$. On the other hand, since E/H varies only between 10 and 25 over the range of materials listed in Table 1, the use of an invariant, representative mean value $\langle(E/H)^{1/8}\rangle = 1.50$ in Eq 5 would lead to errors of no more than 10%. Another potential source of discrepancy lies in the implicit assumption that geometrical similarity is preserved in the indentation pattern from material to material, as reflected in the constancy of the parameters ξ and ψ. We have already pointed out that relaxation effects in the residual contact field can lead to reductions in the ξ term. Systematically low values of ξ will also be manifest in materials that deform by other than a constant-volume process or exhibit plastic pileup at the impression edges [21]. Fused silica and borosilicate glass, which tend to deform by densification [30], fall into this category, thereby explaining the tendency for the data points representing these two materials to lie above the general trend in Fig. 2. Finally, it has been taken as given that the radial crack patterns are always well defined, and in the materials used here this has generally been found to be so. In materials in which the microstructure is comparable in scale with the indentation event, however, the symmetry of the crack pattern can become severely disrupted [8,31], with consequent variations in both ξ and ψ.

It may be argued that the "effective" toughness reckoned from the cutoff position on a master map, while perhaps not an accurate measure of its macroscopically determined counterpart, may nevertheless characterize more closely the response of "natural" flaws. This is certainly likely to be so where the strength-controlling flaw in a component is created by a surface contact event, as in sharp-particle impact or in a machining operation. In this sense the master map approach might well be expected to serve more appropriately as a source of design parameters than the more conventional methods involving large-scale fracture specimens.

Fatigue Curve Slopes—We have noted from Eqs 12 and 13 that the slope of a universal fatigue curve is a measure of the intrinsic susceptibility to slow crack growth. Thus materials with lower values of the crack velocity exponent n, and hence of n' (Eq 9a), will have greater slopes on dynamic fatigue master maps and, conversely, lower (negative) slopes on static fatigue maps.

As with the toughness, certain caution needs to be exercised when using master map data to determine n values. This is because in applying the inverted form of Eq 9a

$$n = \frac{4n'}{3} - \frac{2}{3} \tag{15}$$

it is implicit that certain necessary conditions are met. The most important of these is the proviso $c_0' < c_m$, which we have considered at some length in this work. It is interesting to note that if this proviso is satisfied even the anomalous glasses, which deform by nonvolume-conserving processes, may be analyzed in terms of Eq 15; the fatigue properties are not sensitive to the origin of the residual contact field, as long as this field is of sufficient intensity to generate some precursor crack growth [11]. If such a precursor stage were not to be evident in the failure mechanics the value of the apparent term n' would tend closer to the true n [5,9,12,13]. A second condition that needs to be met is that the flaws should indeed be produced in axial loading; other indentation loading systems, such as linear translation, give rise to flaws which are governed by a transformation equation with coefficients significantly different from those in Eq 15 [12,16].

It is seen in Table 1 that the exponents obtained from this study agree well with the independent determinations for the glasses, but not for the ceramics. The relatively good agreement in the case of the glasses is attributable in part to the model behavior of this class of materials: transparency, isotropy, absence of microstructural complication, and ease in specimen preparation are factors that contribute to this behavior. Also, the n values of the glasses are comparatively low, so fatigue effects show up more strongly. This last point, coupled with a growing realization that conventional testing techniques used to obtain macroscopic velocity data are themselves subject to uncertainty (particularly the double torsion specimen [32]), could account for the discrepancies evident in the data for the ceramics.

Fatigue Curve Intercepts—The intercept terms in the master map representations do not have such a simple interpretation in terms of basic fracture parameters. This is clear from Eq 11; λ_p' is a function of several quantities. Given the fatigue slope and inert strength evaluations, described in the two previous subsections, along with a direct measurement of the critical flaw size, c_m, λ_p' effectively determines the crack velocity coefficient v_0. Because of the compounding of errors (particularly from the n' exponent), determinations of this kind are subject to gross uncertainty. Accordingly, there seems to be little value in trying to retain v_0 as a design parameter, particularly since the λ' terms, which can usually be determined to within 15% from dynamic fatigue data, may be used directly in lifetime formulas. In studies of the basic physics and chemistry of crack growth, of course, v_0 remains a useful coefficient for scientific analysis.

Practical Implications of Master Maps

The major appeal of the master map construction advocated here lies in the provision of a graphic indicator of the intrinsic toughness and fatigue properties of brittle materials. Each material is represented by a universal curve, the relative position of which determines the merit of that material for structural applications. The marked superiority of such materials as silicon carbide and alumina become vividly apparent in the maps of Figs. 5 and 6. Useful distinctions may also be made between materials that cross over within the data range, for example, soda lime glass and lead zircon titanate. On the basis of straight inert strength testing we might reckon the first of these as the stronger material, whereas for applications involving sustained stresses the second would tend to the larger lifetimes. Such crossovers would not be so obvious from the raw fracture mechanics parameters. It will be appreciated that this kind of intercomparison is made on the basis of equivalent flaw sizes: in this respect the indentation method, by virtue of its control over the flaw severity through the contact load, is unique in its capacity for reducing strength data to a common denominator.

In arguing the merits of this approach we do not mean to imply that only the *intrinsic* fracture properties play an important role in the determination of component strengths and lifetimes; the effective sizes of the naturally occurring flaws which ultimately cause failure must also be known. Our procedure, by introducing flaws greater in severity than any of these natural flaws, automatically excludes information concerning the latter from the data. What our scheme effectively allows us to do is to determine the intrinsic parameters in a truly independent manner. All necessary extrinsic flaw parameters should be obtainable from straightforward inert strength tests (run at a single stressing rate), in the form of the usual statistical distribution functions. Lifetime predictions for as-prepared components could then be made without ever having to accumulate vast quantities of fatigue data [2]. In adopting this strategy one needs to keep in mind the strong influence that any persisting residual stress concentrations associated with the original initiation processes (in our case the elastic/plastic deformation) might exert on the subsequent flaw evolution. In the absence of information as to this aspect of flaw characterization, steps should be taken to design conservatively, on the basis of worst-case configurations wherever possible. This last point will be dealt with in greater detail elsewhere in this volume [28].

It has been indicated at several points that the existence of any spurious stresses incurred during the mechanical, chemical, or thermal history of a material would necessitate a third contribution to the starting stress intensity factor in Eq 1, with consequent deviations from the currently determined toughness and fatigue relations. The fact that such deviations were not observed in the materials studied here may be taken as evidence that this potential complication has been avoided successfully. Again, it may be well to emphasize that it is not always simple to confirm the elimination of spurious stresses from surfaces whose strengths are controlled by natural flaws, particularly in materials with typically wide flaw distributions; nor, of course, may we wish to eliminate them, bearing in mind that these stresses are most often compressive.

Finally, a comment may be made concerning the convenience of indentation load as a variable for investigating fundamental flaw size effects. By systematically reducing the load we can produce corresponding smaller flaws, thereby providing a link between macroscopic and microscopic crack behavior. Any change in the nature of the indentation flaw will then become evident as deviations from universal plots, much as just described in relation to the spurious stress influence. In this way it has been possible to demonstrate that indentation flaws in glasses undergo an abrupt transition in properties below a threshold load (corresponding to a flaw size $\approx$ 1 μm): above this threshold the macroscopically determined laws of crack growth remain perfectly valid, regardless of scale, provided the residual contact term is duly accounted for [11]; below the threshold the universal curves no longer apply, and failure becomes dominated by initiation micromechanics [17,33]. The indentation technique should prove similarly useful for studying size effects in ceramics, particularly for polycrystalline materials with relatively coarse microstructures.

Conclusions

1. The indentation-flaw technique provides an attractive route to the evaluation of intrinsic fracture parameters. Coupled with independent determinations of natural flaw distributions, the approach offers the prospect of accurate lifetime predictions with optimum specimen economy.

2. Control over the nature, shape, and, above all, size (by means of the contact load) of the indentation flaw allows for the derivation of a universal fracture formulation. Each material is represented by a single curve, which incorporates the toughness and fatigue properties. Composite plots of these

curves produce master maps, affording a simple graphic format for materials comparisons.

3. The inert strength cutoff on such a master map is a measure of effective material toughness. For "well-behaved" materials this effective toughness is consistent with macroscopically measured K_c values. For cases in which inconsistency is observed, the toughness reckoned from indentation data may provide a more reliable indication of the response of the typical natural flaw.

4. The slope of the fatigue curve on a master map is a measure of the susceptibility of a material to subcritical crack growth. The crack velocity exponent determined from this slope is an apparent value, n', which is converted to the true value, n, by a simple transformation equation.

5. Deviations from universality on a master map indicate an extraneous influence in the fracture mechanics, for example, spurious stress states, microstructure/crack interactions, and threshold size effects.

Acknowledgments

The authors thank T. P. Dabbs for providing raw fracture data on the glasses, and L. Respall and S. J. Mann for help with specimen preparation. Funding was provided by the Australian Research Grants Committee and the U.S. Office of Naval Research (Metallurgy and Ceramics Program).

References

[1] Wiederhorn, S. M., Fuller, E. R., Mandel, J., and Evans, A. G., *Journal of the American Ceramic Society*, Vol. 59, 1976, pp. 404-411.

[2] Ritter, J. E., Bandyopadhyay, N., and Jakus, K., *Ceramic Bulletin*, Vol. 60, 1981, pp. 798-806.

[3] Marshall, D. B. and Lawn, B. R., *Journal of Materials Science*, Vol. 14, 1979, pp. 2001-2012.

[4] Marshall, D. B., Lawn, B. R., and Chantikul, P., *Journal of Materials Science*, Vol. 14, 1979, pp. 2225-2235.

[5] Marshall, D. B. and Lawn, B. R., *Journal of the American Ceramic Society*, Vol. 63, 1980, pp. 532-536.

[6] Chantikul, P., Lawn, B. R., and Marshall, D. B., *Journal of the American Ceramic Society*, Vol. 64, 1981, pp. 322-325.

[7] Anstis, G. R., Chantikul, P., Lawn, B. R., and Marshall, D. B., *Journal of the American Ceramic Society*, Vol. 64, 1981, pp. 533-538.

[8] Chantikul, P., Anstis, G. R., Lawn, B. R., and Marshall, D. B., *Journal of the American Ceramic Society*, Vol. 64, 1981, pp. 539-543.

[9] Lawn, B. R., Marshall, D. B., Anstis, G. R., and Dabbs, T. P., *Journal of Materials Science*, Vol. 16, 1981, pp. 2846-2854.

[10] Cook, R. F., Lawn, B. R., and Anstis, G. R., *Journal of Materials Science*, Vol. 17, 1982, pp. 1108-1116.

[11] Dabbs, T. P., Lawn, B. R., and Kelly, P. L., *Physics and Chemistry of Glasses*, Vol. 23, 1982, pp. 58-66.

[12] Symonds, B. L., Cook, R. F., and Lawn, B. R., *Journal of Materials Science*, Vol. 18, 1983, pp. 1306-1314.

[13] Marshall, D. B. and Lawn, B. R., *Journal of the American Ceramic Society*, Vol. 64, 1981, pp. C6-C7.

[14] Marshall, D. B., Evans, A. G., Khuri-Yakub, B. T., Tien, J. W. and Kino, G. S., *Proceedings of the Royal Society, London*, Vol. A385, 1983, pp. 461-475.

[15] Lawn, B. R., in *Fracture Mechanics of Ceramics*, Vol. 5, R. C. Bradt, A. G. Evans, D. P. H. Hasselman, and F. F. Lange, Eds., Plenum, New York, 1983, pp. 1-25.

[16] Fuller, E. R., Lawn, B. R., and Cook, R. F., *Journal of the American Ceramic Society*, Vol. 66, 1983, pp. 314-321.

[17] Dabbs, T. P. and Lawn, B. R., *Physics and Chemistry of Glasses*, Vol. 23, 1982, pp. 93-97.

[18] Marshall, D. B. and Lawn, B. R., *Journal of the American Ceramic Society*, Vol. 61, 1978, pp. 21-27.

[19] Mould, R. E. and Southwick, R. D., *Journal of the American Ceramic Society*, Vol. 42, 1959, pp. 542-547, 582-592.

[20] Lawn, B. R. and Fuller, E. R., *Journal of Materials Science*, Vol. 10, 1975, pp. 2016-2024.

[21] Lawn, B. R., Evans, A. G., and Marshall, D. B., *Journal of the American Ceramic Society*, Vol. 63, 1980, pp. 574-581.

[22] Wiederhorn, S. M., in *Fracture Mechanics of Ceramics*, Vol. 2, R. C. Bradt, D. P. H. Hasselman, and F. F. Lange, Eds., Plenum, New York, 1974, pp. 613-646.

[23] Wiederhorn, S. M., *Journal of the American Ceramic Society*, Vol. 52, 1969, pp. 99-105.

[24] Ritter, J. E. and Sherbourne, C. L., *Journal of the American Ceramic Society*, Vol. 54, 1971, pp. 601-605.

[25] Cook, R. F., Lawn, B. R., Dabbs, T. P., Reeve, K. D., Ramm, E. J., and Woolfrey, J. L., *Journal of the American Ceramic Society*, Vol. 65, 1982, pp. C172-C173.

[26] McHenry, K. D., Yonushonis, T., and Tressler, R. E., *Journal of the American Ceramic Society*, Vol. 59, 1976, pp. 262-263.

[27] Pletka, B. J. and Wiederhorn, S. M., *Journal of Materials Science*, Vol. 17, 1982, pp. 1247-1268.

[28] Gonzalez, A. C., Multhopp, H., Cook, R. F., Lawn, B. R., and Freiman, S. W., this publication, pp. 43-56.

[29] Lawn, B. R., Marshall, D. B., Chantikul, P., and Anstis, G. R., *Journal of the Australian Ceramic Society*, Vol. 16, 1980, pp. 4-9.

[30] Arora, A., Marshall, D. B., Lawn, B. R., and Swain, M. V., *Journal of Non-Crystalline Solids*, Vol. 31, 1979, pp. 415-428.

[31] Smith, S. S. and Pletka, B. J., in *Fracture Mechanics of Ceramics*, Vol. 6, R. C. Bradt, A. G. Evans, D. P. H. Hasselman, and F. F. Lange, Eds., Plenum, New York, 1983, pp. 189-210.

[32] Pletka, B. J., Fuller, E. R., and Koepke, B. G., in *Fracture Mechanics Applied to Brittle Materials*, ASTM STP 678, American Society for Testing and Materials, Philadelphia, 1978, pp. 19-37.

[33] Dabbs, T. P., Fairbanks, C. J., and Lawn, B. R., this publication, pp. 142-153.

Armando C. Gonzalez,[1] *Heidi Multhopp,*[1] *Robert F. Cook,*[1]
Brian R. Lawn,[1] *and Stephen W. Freiman*[1]

Fatigue Properties of Ceramics with Natural and Controlled Flaws: A Study on Alumina

REFERENCE: Gonzalez, A. C., Multhopp, H., Cook, R. F., Lawn, B. R., and Freiman, S. W., **"Fatigue Properties of Ceramics with Natural and Controlled Flaws: A Study on Alumina,"** *Methods for Assessing the Structural Reliability of Brittle Materials, ASTM STP 844,* S. W. Freiman and C. M. Hudson, Eds., American Society for Testing and Materials, Philadelphia, 1984, pp. 43–56.

ABSTRACT: A systematic study has been made of the fatigue properties of an as-fired polycrystalline alumina containing either "natural" (sawing damage) or indentation-induced (Vickers) strength-controlling flaws. All fatigue strengths were measured in four-point bending in water. The study is presented in three steps: first, comparative Weibull analyses are made of inert strength data for the two flaw types, both to demonstrate the reduction in scatter that attends the indentation method and to characterize the flaw distributions for the as-sawn surfaces; next, fatigue data are taken on indented surfaces to determine relatively accurate fracture parameters for the alumina and to confirm that constant stressing rate tests can be used as a base for predicting the response in static loading; finally, the results from the two previous, independent steps are combined to generate lifetime responses for the surfaces with natural flaws, and fatigue data taken on such surfaces are used to evaluate these predictions. It is emphasized that residual stresses around the critical flaws (associated either with the preceding contact events responsible for creating the flaws or with extraneous processing, preparation, or service conditions) can play a crucial role in the fracture mechanics. Notwithstanding this complication, the present approach offers a new design philosophy, with the potential for predicting responses relating to flaws generated after, as well as before, any laboratory screening tests.

KEY WORDS: alumina, fatigue, indentation flaw, lifetime prediction, residual stresses, strength testing, structural reliability, brittle materials

In the preceding paper [*1*] a case was made for using indentation flaws to investigate the fracture properties of candidate materials for structural applications. The indentation method allows for complete control over the forces used

[1] Engineer, guest student, graduate student, physicist, and engineer, respectively, Center for Materials Science, National Bureau of Standards, Washington, D.C. 20234.

to generate the critical flaws, provides knowledge of the local stress state of these flaws prior to strength testing, and reduces the scatter in the ensuing failure stresses. Most important, it divides the general strength problem into its two constituent parts, facilitating truly independent determinations of intrinsic material parameters and extrinsic flaw distribution characteristics. This opens the way to a new approach to design, whereby much of the empiricism and statistical data handling associated with conventional strength testing might be avoided.

In this study we demonstrate the approach on a commercial alumina. Alumina was chosen because of its widespread use as a structural ceramic, its availability in large quantities, its relatively simple microstructure, and, above all, its well-documented susceptibility to slow crack growth. This last point is a key one, for it highlights the variability that can bedevil fracture mechanics measurements in ceramics; evaluations of the crack velocity exponent n, using both macroscopic crack specimens [2-16][2] and fatigue strength tests [6-8],[2] lie anywhere between 30 and 90. Part of this variability is no doubt attributable to differences in the source materials. However, the increasing recognition that most techniques in current use for monitoring crack growth are subject to systematic error [9], coupled with the strong influence that any residual stress fields around the critical flaws have on the slopes of fatigue curves [1], can also account for significant discrepancies. The recent comparative study by Pletka and Wiederhorn of double torsion and strength tests on common-source aluminas and other ceramics suggests that such discrepancies could easily exceed a factor of three [6]. There would appear to be a need for greater awareness of the oversimplistic assumptions that are implicit in our present descriptions of crack growth laws, at both the macroscopic and microscopic levels.

The central aim of the present work is to characterize the strength properties of alumina specimens containing "controlled" flaws in order to optimize the amount of testing that must be carried out on similar specimens with "natural" flaws. More specifically, it is intended that crack growth parameters for the alumina should be obtained from dynamic fatigue results on indented surfaces, and flaw distribution parameters from independent inert strength tests on as-prepared surfaces, enabling the two vital elements of the lifetime prediction problem to be treated separately. Predictions made using this approach will be tested against representative fatigue data from the latter, natural surfaces.

Experimental Procedure

Preparation of Specimens with Different Flaw Types

The aluminum oxide used in this study was a roll-compacted, sintered substrate material with 4% additive component (AD96, Coors Porcelain, Colo-

---

[2]B. J. Koepke, Honeywell, unpublished work, 1980.

rado), having an average grain size $\approx 10\ \mu$m. It was obtained as plates 1.3 mm thick in its as-fired state and was diamond-sawn into strips 30 mm long and 5 mm wide. Inspection of these strips in the optical microscope revealed chipping damage at the edges. One group of specimens was immediately selected out, at random, and set aside for testing in the as-received state.

The remaining specimens were used for controlled-flaw testing. Each member of this group was indented at a face center with a Vickers diamond pyramid, care being taken to orient the impression diagonals parallel to the specimen edges. For this purpose, a standard load of $P = 5$ N was chosen; this represented a compromise between the requirements that the radial cracks extending from the impression corners should be sufficiently large in comparison with the scale of the impression itself, and yet sufficiently small in comparison with the specimen thickness [10]. All indentations were made in air and were allowed to sit ≈ 1 h prior to strength testing. Optical and scanning electron microscopical examination of representative examples on surfaces prepolished through 3-μm diamond paste showed that the crack patterns thus produced were not generally of the ideal radial geometry, because of microstructural complications [11], as is evident in the photomicrograph of Fig. 1. The inden-

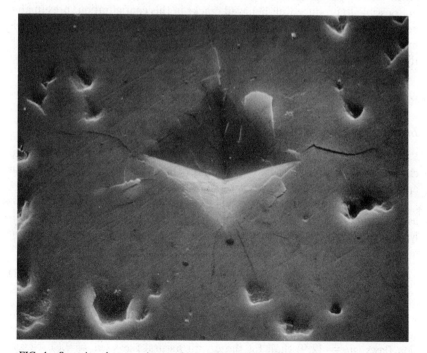

FIG. 1—*Scanning electron micrograph of a Vickers indentation flaw in alumina. Note the irregular nature of the radial cracking about the hardness impression. Indentation load* P = 5 N; *width of field 150 μm.*

tations were nevertheless sufficiently well formed to allow hardness determinations and, to a lesser extent, crack size measurements to be made.

Strength Testing

The alumina bars prepared as previously described were loaded to failure in four-point flexure, outer span 27 mm and inner span 9 mm, with the surfaces containing the indentation flaws on the tension side. Inert strengths were measured in nitrogen gas or silicone oil, fatigue strengths in water. The inert strength and dynamic fatigue tests were run using a crosshead loading machine, the former at the fastest available rate. Breaking loads were measured by strain-gage and piezoelectric cells [10]. For the static fatigue tests the load was applied pneumatically,[3] with a nominal rise time of 8 s and a maximum fluctuation of 1% at hold. Simple beam theory was used to evaluate the stresses from the recorded loads.

All the broken test pieces were examined by optical microscopy to confirm the sources of failure. As expected, those specimens with controlled flaws broke from the indentation sites and those without from the as-sawn edges.

Efforts were also made to run double torsion tests on the alumina, to obtain crack velocity parameters as a check on the strength analysis. However, it was not possible to produce well-behaved cracks in this configuration, presumably because of instabilities in the propagation [6]. Double-cantilever beam specimens were also unsatisfactory, because of the difficulty in locating the crack tips.

Results

Inert Strength Tests

Inert strength tests were run to determine flaw statistical parameters, to check for spurious preexisting stresses in the specimen surfaces, and to obtain appropriate toughness parameters for later fatigue analysis.

The first runs were made on specimens from each of the two groups, that is, as-sawn and indented. The data from these runs, shown in Fig. 2, were analyzed in accordance with the usual two-parameter Weibull probability function

$$F = 1 - \exp\left[-\left(\frac{\sigma_m}{\sigma_0}\right)\right]^m \tag{1}$$

where σ_m is the inert strength and m and σ_0 are adjustable parameters. It is seen that the spread in results is indeed smaller for the surfaces with indentation flaws ($m = 12.9$) than for those with natural flaws ($m = 9.8$). Never-

[3] A. C. Gonzalez and S. W. Freiman, National Bureau of Standards, unpublished work, 1983.

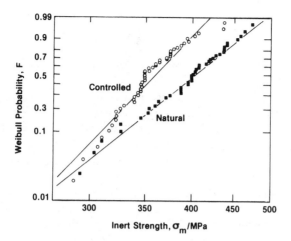

FIG. 2—*Weibull plot of inert strengths on alumina surfaces with natural (sawing damage) and controlled (5 N Vickers indentation) flaws.*

theless, this spread in the former case is by no means insignificant, consistent with the inherent variability in the crack pattern of Fig. 1.

The next runs were made on indented specimens as a function of the contact load, P. Figure 3 shows the results. The data points represent strengths at $50 \pm 32\%$ Weibull failure probability (equivalent to standard deviation limits for a normal distribution) for at least ten specimens per load, and the straight line is a best fit of slope $-\frac{1}{3}$, in logarithmic coordinates, from which we obtain $\sigma_m P^{1/3} = 590 \pm 47$ MPa N$^{1/3}$ (mean and standard deviation). The constancy of this quantity over the load range covered is an indication of the absence of

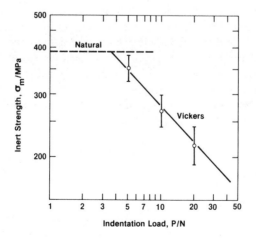

FIG. 3—*Inert strength of alumina as a function of Vickers indentation load.*

prepresent stresses in the as-fired surfaces [1,12]. It is noted that the as-sawn strength level in Fig. 3 corresponds to an effective indentation load of 3.3 ± 1.4 N.

Some additional tests were made on prepolished surfaces containing 5-N "dummy" indentations [1,10]. In these tests, on ten specimens, failure occurred from one of three near-identical contact sites located along the specimen within the inner span, leaving two dummies intact for the determination of the important crack dimensions. Thus, measurements of the set of radial cracks parallel to the tensile direction gave the initial crack dimension of $c_0' = 28 ± 4$ μm, while those of the perpendicular set gave the critical dimension of $c_m = 33 ± 5$ μm. The crucial proviso for validity of the fracture mechanics formulation in Ref 1, that is, $c_0' \le c_m$ is therefore satisfied.

With the underlying basis of the equilibrium fracture description thereby established, we may insert the value of $\sigma_m P^{1/3}$ obtained previously, together with $H = 15.5 ± 1.0$ GPa measured directly from the hardness impressions and $E = 303$ GPa specified by the manufacturer, into the expression for toughness [1,13]

$$K_c = \eta \left(\frac{E}{H} \right)^{1/8} (\sigma_m P^{1/3})^{3/4} \qquad (2)$$

where $\eta = 0.59$. This gives $K_c = 3.2 ± 0.2$ MPa m$^{1/2}$, which may be compared with the value 3.31 ± 0.07 obtained by other workers [14] on similar material using a chevron-notched rod technique.

Dynamic and Static Fatigue of Specimens with Controlled Flaws

Dynamic and static fatigue data were collected on the alumina specimens with standard 5 N indentation flaws, with the purpose of testing the theoretically predicted interrelationships between the two stressing modes.

Figure 4 shows the dynamic fatigue results. The data points are 50 ± 32% Weibull evaluations of the strengths, σ_f, for at least ten specimens at each of the prescribed stressing rates, $\dot{\sigma}_a$. It is immediately evident that the fatigue strengths are substantially less than the inert strength level, even at the fastest stressing rates. The straight line is a best fit to all individual test results, in accordance with the prediction [1,15-16]

$$\sigma_f = (\lambda' \dot{\sigma}_a)^{1/(n' + 1)} \qquad (3)$$

Bearing in mind the precursor growth stage apparent in the equilibrium failure mechanics referred to in the previous subsection (that is, $c_0' < c_m$), it is important to emphasize that the slope and intercept terms, n' and λ', relate to *apparent* crack velocity parameters. Appropriate transformation equations for converting these to corresponding *true* parameters obtain from the

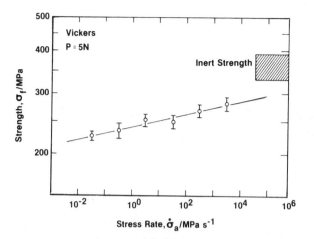

FIG. 4—*Dynamic fatigue of alumina in water, for specimens with controlled flaws.*

modified stress intensity factors for point flaws with residual contact terms incorporated [1,16]

$$n' = \frac{3n}{4} + \frac{1}{2} \tag{4a}$$

$$\lambda' = \frac{(2\pi n')^{1/2}\sigma_m^{n'}c_m}{v_0} \tag{4b}$$

where n and v_0 are exponent and coefficient, respectively, in the usual power-law crack velocity relation (Eq 6 in Ref 1). Thus, from the data analysis we obtain $n' = 54.9 \pm 4.9$, which converts, in Eq 4a, to $n = 72.5 \pm 6.5$; also, log $\lambda' = 133 \pm 12$ (in the units used in Fig. 4), which, in conjunction with the inert strength data from the preceding subsection, yields log $v_0 = 4.0 \pm 0.5$ (velocity in metres per second).

The corresponding results for the static fatigue tests are shown in Fig. 5. In this case the data are plotted as *median* values of the times to failure, t_f, over ten tests at each of the prescribed holding stresses, σ_A, to accommodate null tests in which the specimens either broke during the loading ramp or survived the two-week cutoff. The straight lines are predictions using the static analogue of Eq 3 [1,16], that is

$$t_f = \frac{\lambda'}{(n'+1)\sigma_A^{n'}} \tag{5}$$

where the terms n' and λ' have the same values as previously. As discussed in Ref 1, the procedure is equivalent to inverting the dynamic fatigue curve in ac-

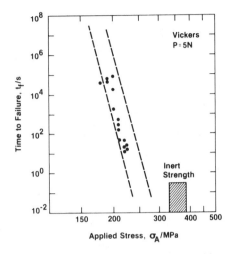

FIG. 5—*Static fatigue of alumina in water, for specimens with controlled flaws.*

cordance with the relation $t_f = \sigma_f/\dot{\sigma}_a$, identifying σ_f with σ_A, and translating the intercept through $n' + 1$ in logarithmic space. In the spirit of this description we have generated 17 and 83% failure probability limits directly from the corresponding Weibull band for the inert strengths in Fig. 5. The level of agreement between data and predictions in this figure may be taken as a measure of the confidence with which we might use the fatigue equations to analyze the response of less well defined flaws.

Lifetimes of Specimens with Natural Flaws

In this part of the study *a priori* predictions were made of the lifetime characteristics for surfaces with the natural (sawing damage) flaws, using the results of the preceding subsections. Fatigue data were then taken on such surfaces as a check against these predictions.

In adopting this course, we find ourselves confronted immediately by an apparent obstacle, namely, our lack of foreknowledge of the flaw characteristics. If we could assume that the natural flaws were to behave in essentially the same way as the Vickers-induced radial cracks, the procedure would be straightforward enough. Then, one could make use of the "effective" load evaluated at the intersection point of the indentation line, $\sigma_m P^{1/3} = $ constant, with the inert strength level in Fig. 3 to characterize the flaw severity. The appropriate lifetime relation would follow directly from Eq 5, using the same slope parameter, n', as determined for the indentation flaws but with a load-adjusted intercept parameter [1]

$$\lambda' = \frac{\lambda_p'}{p^{(n' - 2)/3}} \tag{6}$$

where λ_P' is a modified, load-independent term, also to be evaluated from the indentation fatigue data. This prediction is plotted as the solid line in Fig. 6. As in Fig. 5, failure probability limits may be generated directly from the plotted 50 ± 32% Weibull band for the inert strengths, but these are omitted from the present plot for the sake of clarity.

Unfortunately, the assumption that the strength properties of real materials may be described in terms of ideal point contact flaws does not always hold to good approximation [1]. If the past history of the controlling flaws is such that residual driving forces do indeed persist to stabilize the initial crack growth, but the flaw has essentially *linear* rather than point geometry, the mechanics will reflect the same kind of stress augmentation, but with even greater intensity [17]. Or, if for some reason the residual influence is diminished to an insignificant level, the mechanics will tend closer to those for Griffith flaws (that is, zero residual stress) [18,19]. In either case, the procedure for generating a lifetime prediction remains much the same as before, in that Eq 5 may be retained as the basic starting formula but with the slope term n' in Eq 4a and the intercept term λ' in Eq 4b replaced by appropriate analogues [16]. Expressions for these replacement terms are given in the Appendix; suffice it to say here that evaluations may still be made from the independently obtained dynamic fatigue data on the indented control specimens and inert strength data on the actual specimens with natural flaws. The predictions for these alternative, extreme flaw types are plotted as the broken lines in Fig. 6.

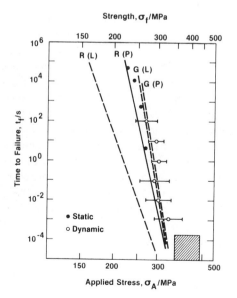

FIG. 6—*Lifetime diagram for alumina in water, for surfaces with natural flaws. The lines are predictions based on indentation-calibrated fracture mechanics formulas: R refers to flaws with residual stress, G to conventional Griffith flaws; P and L denote flaws with point and line geometry, respectively. The points are confirmatory static and dynamic fatigue data.*

These various predictions for the natural surfaces may be compared with the data points from the confirmatory fatigue tests. Initially, it was intended that all such data should be obtained directly from tests at constant applied stress, but the problems with loading-ramp failures and runouts referred to in connection with Fig. 5, magnified in the case of natural surfaces by the wider spread in flaw severity (that is, lower Weibull modulus), imposed severe limitations on specimen and time economy. Therefore, supplementary data were collected in the constant stressing rate mode, using the inversion and translation operation which interrelates Eqs 3 and 5 to evaluate equivalent lifetimes. A difficulty here, of course, is that this conversion operation is contingent on the quantity $n' + 1$, which we cannot specify *a priori*. In our case, we have obtained a working evaluation by regressing on the pooled dynamic fatigue results, in accordance with Eq 3. Accordingly, the data points in Fig. 6 represent median values for 20 to 40 tests at each static holding stress or 50 $\pm$ 32% probability bounds for 10 to 15 tests at each predetermined dynamic stress rate.

Discussion

We have presented results of a strength study on alumina surfaces with both controlled and natural flaws in the context of lifetime design. Currently, it is widely accepted that the most reliable route to this end is via the exclusive and extensive testing of specimens with the same preparation as that of the structural component, regarding Eqs 3 and 5 as empirical relations to be used in conjunction with statistically determined flaw distributions [20]. We have argued for an alternative philosophy, in which inert strength tests on as-prepared surfaces are retained to determine the flaw distributions, but independent tests are run on indented surfaces to evaluate toughness and crack velocity parameters. The most apparent advantage of this approach is a substantial reduction in the uncertainty in the intrinsic, material component of the strength formulation, so that many fewer specimens should need to be broken to attain a specified tolerance in predicted lifetimes.

The one major obstacle we face in adopting this alternative course is the general inability to predetermine the true nature of the strength-controlling flaw in any prospective structural component. We have seen in Fig. 6 that the presence of residual stresses about the flaw center and the geometrical aspect of this flaw configuration can be decisive factors in lifetime response. In the present tests on as-sawn specimens the results would appear to indicate a relatively minor role for these factors. This is not altogether unreasonable, for, although diamond-sawing is a contact-related process, the basic removal mechanism is one of "lateral-crack" chipping [21], and such chipping modes can greatly relieve the residual contact fields [18]. In principle, the extent of such relief mechanisms may be quantified by comparing strength values before and after a full anneal treatment of the natural surfaces, as has been done in glass [18,19,22]. With many ceramics, however, annealing is impractical, in which case the prudent designer would presume that residual stress components remain fully active. Indeed, in certain cases, such as with machined surfaces, it might well be advisable to adopt the ultraconservative path and work on the assumption that the flaws also have essentially linear geometry. There is clearly the prospect of overdesign with this approach, which may be unacceptable in applications where the limits of material performance are an absolute necessity. Any decision to design on a less conservative basis, on the other hand, should be backed up with confirmatory fatigue data, such as in Fig. 6. Then, of course, we shall have had to revert, at least in part, to precisely the kind of testing we have sought to supplant in the first place.

With due acknowledgment of the complication just discussed, we may now reinforce our case for the controlled-flaw procedure advocated in this work by emphasizing some of the unique advantages which attend the broad field of indentation fracture mechanics [23–25]. Most important, the approach offers, with its physical insight into the underlying micromechanics of flaw development, the prospect of accommodating changes in the flaw characteristics *subsequent to the laboratory screening tests* within the design specifications. Such changes can be particularly dangerous if they are associated with the spontaneous initiation of *new* flaws, due, for example, to interactions with a hostile mechanical [24,26,27] or chemical [28] service environment. Under these conditions any amount of laboratory testing on as-prepared surfaces would be totally useless if the new flaws were to be dominant. However, provided that the potential service environment is specifiable, indentation fracture mechanics provides us with the facility for estimating an equivalent indentation load for any such flaw; in a particle erosion field, for instance, the load is readily calculable in terms of the incident particle energy and quasi-static component hardness [27]. The problem is thereby reduced to the level of the prepresent natural flaw, whence Eq 6 may be invoked, as before, to obtain a lifetime prediction from Eq 5.

Another distinctive advantage of the indentation flaw method is that one can check routinely for spurious stresses in the as-prepared surfaces. The presence of such stresses becomes manifest as a breakdown in the fracture mechanics formalism used earlier in this work, most conveniently in the inert strength response as a departure from the load independence of the quantity $\sigma_m P^{1/3}$ [1]. (Indeed, *quantitative* information on surface compression stresses has been determined in this manner for tempered glasses [29,30]). Insofar as the lifetime predictions in Fig. 6 are concerned, the effect of a superposed spurious stress may be regarded in terms of an appropriate displacement of all plotted points, in absolute terms, along the horizontal axis, thereby introducing a greater or lesser degree of curvature in the logarithmic representation [15]. This curvature may pass unnoticed in tests on natural surfaces, depending on the scatter in data and range in failure times covered, yet lead to significant discrepancies in long-term extrapolations.

Acknowledgments

The authors thank Dan Briggs of Coors Porcelain Co. for supplying the alumina plates used in this study. Funding was provided by the U.S. Office of Naval Research (Metallurgy and Ceramics Program).

APPENDIX

In interpreting the fatigue results for specimens with natural flaws it was indicated that Eqs 3 and 5 may be retained as the basis for analysis, provided that the quantities n' in Eq 4a and λ' in Eq 4b for point flaws with residual stresses are suitably replaced to match the specific flaw characteristics. Reference is made to the paper by Fuller et al [16] for details.

One of the distinctions we shall be required to make in effecting these conversions is that between geometrical factors for the point and line configurations. The relationship between inert strength, σ_m, and critical crack size, c_m, for the standard point indentations

$$\sigma_m = \frac{3K_c}{4\Psi_p c_m^{1/2}} \tag{7}$$

provides us with the means for doing this; here Ψ_p is a dimensionless factor to be evaluated from the experimental data. An equivalent evaluation for line flaws may then be made purely on theoretical grounds, using an appropriate "modification" relation [31]

$$\Psi_\ell = \frac{\pi}{2}\Psi_p \tag{8}$$

Consider now the case of linear flaws with fully persistent residual stresses. The replacement quantities in the fatigue equations are

$$n'' = \frac{n}{2} + 1 \tag{9a}$$

$$\lambda'' = \frac{(4\pi n'')^{1/2}\sigma_m^{n''} c_m}{v_0} \tag{9b}$$

where it is understood that σ_m and c_m now pertain to measurements on the natural surfaces. Since crack sizes are not readily measured for failures from natural flaws, it is convenient to eliminate c_m from Eq 9b using the line-flaw analogue of Eq 7

$$\sigma_m = \frac{K_c}{2\Psi_\ell c_m^{1/2}} \tag{10}$$

Thus, given the calibrated values of Ψ_ℓ and K_c from the standard indentation tests, we are left with the natural inert strength as the controlling variable in Eq 9b. Equation 9 may then be coupled with its indentation-flaw counterpart, Eq 3 in the text, to eliminate n and v_0, thus completing the conversion operation.

For flaws with zero residual stress, conventional theory applies. The replacement quantities are

$$n_0 = n \qquad (11a)$$

$$\lambda_0 = \left[\frac{2}{(n-2)} \right] \frac{\sigma_0'^n c_0'}{\nu_0} \qquad (11b)$$

where σ_0' is the inert strength conjugate to the initial flaw size c_0'. In this case the apparent and true crack velocity exponents are identical. The crack size may be eliminated through the familiar inert strength relation

$$\sigma_0' = \frac{K_c}{\Psi c_0'^{1/2}} \qquad (12)$$

where Ψ identifies with the geometrical factor for point or line flaws, as appropriate. Thereafter, the procedure is the same as that outlined in the previous sample.

References

[1] Cook, R. F. and Lawn, B. R., in this publication, pp. 22-42.
[2] Freiman, S. W., McKinney, K. R., and Smith, H. L., in *Fracture Mechanics of Ceramics*, Vol. 2, R. C. Bradt, D. P. H. Hasselman, and F. F. Lange, Eds., Plenum, New York, 1974, pp. 659-676.
[3] Evans, A. G., Linzer, M., and Russell, L. R., *Materials Science and Engineering*, Vol. 15, 1974, pp. 253-261.
[4] Bansal, G. K. and Duckworth, W. H., *Journal of Materials Science*, Vol. 13, 1978, pp. 215-216.
[5] Ferber, M. K. and Brown, S. D., *Journal of American Ceramic Society*, Vol. 63, 1980, pp. 424-429.
[6] Pletka, B. J. and Wiederhorn, S. M., *Journal of Materials Science*, Vol. 17, 1982, pp. 1247-1268.
[7] Rockar, E. M. and Pletka, B. J., in *Fracture Mechanics of Ceramics*, Vol. 4, R. C. Bradt, D. P. H. Hasselman, and F. F. Lange, Eds., Plenum, New York, 1978, pp. 725-735.
[8] Ritter, J. E. and Humenik, J. N., *Journal of Materials Science*, Vol. 14, 1979, pp. 626-632.
[9] Freiman, S. W., in *Fracture Mechanics of Ceramics*, Vol. 6, R. C. Bradt, A. G. Evans, D. P. H. Hasselman, and F. F. Lange, Eds., Plenum, New York, 1983, pp. 27-45.
[10] Cook, R. F., Lawn, B. R., and Anstis, G. R., *Journal of Materials Science*, Vol. 17, 1982, pp. 1108-1116.
[11] Anstis, G. R., Chantikul, P., Marshall, D. B., and Lawn, B. R., *Journal of the American Ceramic Society*, Vol. 64, 1981, pp. 534-538.
[12] Marshall, D. B. and Lawn, B. R., *Journal of Materials Science*, Vol. 14, 1979, pp. 2001-2012.
[13] Chantikul, P., Anstis, G. R., Lawn, B. R., and Marshall, D. B., *Journal of the American Ceramic Society*, Vol. 64, 1981, pp. 539-543.
[14] Barker, L. M., in *Fracture Mechanics of Ceramics*, Vol. 3, R. C. Bradt, D. P. H. Hasselman, and F. F. Lange, Eds., Plenum, New York, 1978, pp. 483-494.
[15] Lawn, B. R., Marshall, D. B., Anstis, G. R., and Dabbs, T. P., *Journal of Materials Science*, Vol. 16, 1981, pp. 2846-2854.
[16] Fuller, E. R., Lawn, B. R., and Cook, R. F., *Journal of the American Ceramic Society*, Vol. 66, 1983, pp. 314-321.
[17] Symonds, B. L., Cook, R. F., and Lawn, B. R., *Journal of Materials Science*, Vol. 18, 1983, pp. 1306-1314.
[18] Marshall, D. B. and Lawn, B. R., *Journal of the American Ceramic Society*, Vol. 63, 1980, pp. 532-536.

[19] Chantikul, P., Lawn, B. R., and Marshall, D. B., *Journal of the American Ceramic Society*, Vol. 64, 1981, pp. 322–325.

[20] Ritter, J. E., in *Fracture Mechanics of Ceramics*, Vol. 5, R. C. Bradt, A. G. Evans, D. P. H. Hasselman, and F. F. Lange, Eds., Plenum, New York, 1982, pp. 227–251.

[21] Marshall, D. B., Lawn, B. R., and Evans, A. G., *Journal of the American Ceramic Society*, Vol. 65, 1982, pp. 561–566.

[22] Marshall, D. B. and Lawn, B. R., *Journal of the American Ceramic Society*, Vol. 64, 1981, pp. C6–C7.

[23] Lawn, B. R. and Wilshaw, T. R., *Journal of Materials Science*, Vol. 10, 1975, pp. 1049–1081.

[24] Lawn, B. R., in *Fracture Mechanics of Ceramics*, Vol. 3, R. C. Bradt, D. P. H. Hasselman, and F. F. Lange, Eds., Plenum, New York, 1978, pp. 205–229.

[25] Lawn, B. R., in *Fracture Mechanics of Ceramics*, Vol. 5, R. C. Bradt, A. G. Evans, D. P. H. Hasselman, and F. F. Lange, Eds., Plenum, New York, 1983, pp. 1–25.

[26] Hockey, B. J., Wiederhorn, S. M. and Johnson, H., in *Fracture Mechanics of Ceramics*, Vol. 3, R. C. Bradt, D. P. H. Hasselman, and F. F. Lange, Eds., Plenum, New York, 1978, pp. 379–402.

[27] Wiederhorn, S. M. and Lawn, B. R., *Journal of the American Ceramic Society*, Vol. 62, 1979, pp. 66–70.

[28] Dabbs, T. P., Fairbanks, C. J., and Lawn, B. R., in this publication, pp. 142–153.

[29] Marshall, D. B. and Lawn, B. R., *Journal of the American Ceramic Society*, Vol. 61, 1978, pp. 21–27.

[30] Chantikul, P., Marshall, D. B., Lawn, B. R., and Drexhage, M. G., *Journal of the American Ceramic Society*, Vol. 62, 1979, pp. 551–555.

[31] Lawn, B. R. and Wilshaw, T. R., *Fracture of Brittle Solids*, Cambridge University Press, London, 1975, Chapter 3.

D. K. Shetty,[1] A. R. Rosenfield,[1] and W. H. Duckworth[1]

Statistical Analysis of Size and Stress State Effects on the Strength of an Alumina Ceramic

REFERENCE: Shetty, D. K., Rosenfield, A. R., and Duckworth, W. H., "**Statistical Analysis of Size and Stress State Effects on the Strength of an Alumina Ceramic,**" *Methods for Assessing the Structural Reliability of Brittle Materials, ASTM STP 844*, S. W. Freiman and C. M. Hudson, Eds., American Society for Testing and Materials, Philadelphia, 1984, pp. 57–80.

ABSTRACT: Fracture strengths of an alumina ceramic, surface-finished to obtain an isotropic flaw population, were evaluated in inert and water environments in three qualified tests—four-point and three-point bend tests and a biaxial flexure test featuring uniform-pressure loading of disk specimens. The resulting strengths were examined in terms of a statistical fracture theory that employed fracture criteria derived from fracture mechanics theory and supported by reported experimental results on the fracture from angled surface cracks from indentation. The four-point and three-point bend tests gave fracture strengths, in both inert and water environments, that were consistent with the size effect predicted by the statistical fracture theory. The stress state effect on strength, as indicated by the results of the biaxial test, was predicted by the theory only for the tests in water. The inert biaxial strengths were significantly greater than the statistical theory predictions, irrespective of the fracture criteria employed. The differences in the biaxial strengths predicted from the theory, on the bases of a normal stress (Mode I) and a combined mode fracture criteria, were small because of the relatively small values of the strength scatter in each test. The observed trend in the biaxial fracture strengths in relation to the uniaxial strengths with the change in the test environment suggested more severe strength degradation from slow-crack growth in the biaxial stress state than that in the uniaxial stress state. Implications of this result with respect to long-term reliability of brittle structural components are discussed.

KEY WORDS: brittle materials, structural reliability, fracture strengths, fracture criteria, size effect, mixed-mode fracture, stress state effect, strength distributions, subcritical crack growth, residual stresses, statistical analysis, crack branching

Nomenclature

σ_r Radial stress in the disk specimen at a radial position, r

σ_t Tangential stress in the disk specimen at radial position, r

[1]Principal research scientist, research leader, and member of the research council, respectively, Battelle-Columbus Laboratories, Columbus, Ohio 43201.

σ_u Maximum stress at the disk center

r_1 Disk specimen support radius

r_2 Disk specimen radius

α, β Nondimensional parameters describing the stress gradients in the disk specimens

p Applied pressure in the biaxial disk test

ν Poisson's ratio

E Young's modulus

V Subcritical crack growth rate

K_I Applied stress intensity factor

K_{Ic} Critical stress intensity factor

V_0, n Parameters in the empirical subcritical crack growth equation

F Cumulative fracture probability

Ω A solid angle within which crack normals should lie for fracture to occur

$N(\sigma_c)$ A function representing number of surface cracks per unit area that will cause fracture when remote tension normal to the crack plane is $\leq \sigma_c$

$\bar{k}, m$ Flaw size distribution parameters

σ_N Tensile stress normal to a crack

τ Shear stress parallel to a crack

σ_f Maximum tensile stress in the four-point bend specimen

σ_t Maximum tensile stress in the three-point bend specimen

L Support span length

a Loading span length for four-point bend specimens

b Bend specimen thickness

t Bend specimen width

Γ Gamma function

i Ranking number of a specimen

S Sample size

Y A fracture mechanics parameter that incorporates crack shape, stress gradient, and free surface effects on stress intensity

σ_b Tangential stress at the crack branching point

a_b Crack length at branching

σ_R, A_b Empirical parameters in the crack branching relation

Fracture strengths of brittle materials are usually evaluated only in simple uniaxial tests, such as three-point or four-point bend tests, whereas in structural applications the materials are subjected to complex multiaxial stress states. The relationship between conventional bend strengths and the fracture stresses of a component in general service loading thus becomes critical to ensuring the reliability of components, because it directly determines either the allowable design stress for an acceptable probability of failure or the stress level in proof testing for a selected component rejection rate. The strength relations must suitably account for the effects of flaw populations (size effect) as well as the effects of interactions of the principal stresses in general loading (stress state effect).

A number of studies on ceramics have examined the relationships between strengths in different stress states. Duckworth and Rosenfield [1] reviewed the results reported up until 1978. It was found that the results on ceramics in general did not provide a consistent picture of the stress state effects on strength. A number of studies reported a weakening effect of both tension-tension and tension-compression stress states, and others indicated an opposite effect. Some of the anomaly in the data appears to be due to the effects of anisotropic distributions of surface flaws. This resulted in different flaw populations being sampled in the different stress state tests. The studies of Babel and Sines [2] also suggested that the nature of the strength-controlling flaws in ceramics, for example, volume flaws like pores versus cracklike surface flaws, was influential in determining the effects of stress states on strength.

The increasing application of ceramics as structural components has renewed studies of stress state effects on ceramic strength. Giovan and Sines [3] compared the fracture strengths of an alumina ceramic tested in four-point bend and a ring-on-ring test. For specimens with a lapped-surface finish and for a group of longitudinally ground specimens, the mean biaxial strength was less than the mean four-point bend strength in accord with a statistical fracture theory due to Batdorf [4]. The classic Weibull theory [5] overestimated the weakening effect in the biaxial tension test. But for a second group of ground specimens, Giovan and Sines [3] found the uniaxial strength to be less than the biaxial strengths, particularly for the transverse-ground specimens.

The present authors examined the uniaxial and biaxial fracture strengths of a glass ceramic under controlled conditions of surface finish, environment, and stressing rate [6]. The glass ceramic exhibited unusually high uniformity of fracture stresses and a resultant lack of size-dependence of strength in uniaxial bend tests. The fracture strength of the glass ceramic in a biaxial flexure test was significantly greater than the uniaxial strengths when all the tests were done in an inert environment (dry nitrogen gas). This behavior was termed biaxial strengthening. In a water environment and under conditions of low stressing rates the fracture strengths in the uniaxial and the biaxial tests were nearly equal. Thus, this study showed that the strength degradation due to subcritical crack growth in the glass ceramic was more pronounced in a biaxial stress state. The more recent experimental results of Pletka and Wiederhorn [7] confirm this finding; they observed a greater stressing rate dependence of alumina ceramic strengths in a biaxial flexure test than in a uniaxial flexure test. Petrovic and Stout recently compared the fracture strengths of an alumina ceramic in combined tension-torsion tests with the strengths obtained in uniaxial tension [8]. The tension-torsion strengths were greater than the uniaxial strengths. This result was qualitatively consistent with the predictions of the statistical fracture theories, but the magnitude of the increase in strength in the combined tension-torsion test was greater than that predicted by the classic Weibull theory [9].

The present paper is based on the continuing research at Battelle's Columbus Laboratories on the stress state effects on strengths of ceramics. A new biaxial flexure test featuring axisymmetric pressure loading of disks and a statistical analysis of preliminary fracture strength data obtained on a commercial alumina ceramic were discussed in a recent paper by the present authors [10]. Weibull's statistical theory of strength [5] adequately accounted for the specimen size effect by accurately predicting three-point bend strengths from parameters derived from four-point bend strength data. Neither the Weibull theory nor a modification of it [11], however, was completely satisfactory in predicting biaxial strengths. Specifically, biaxial strengths predicted from the uniaxial strength parameters were significantly less than the measured strengths. This result was qualitatively similar to our earlier finding in the glass ceramic [6]. Additional fracture strength data for the alumina ceramic obtained in both inert and water environments are presented and discussed in this paper. Specimen size and stress state effects on the fracture strength are examined in terms of a generalized statistical fracture theory [4] incorporating two different fracture criteria. Implications of the experimental and analytical results with respect to the current methods of assessing structural reliability of brittle materials for long-term applications in slow-crack-growth environments are discussed.

Experimental Procedures

Strength Tests

Two uniaxial strength tests, four-point and three-point bend tests, and a biaxial strength test, a disk flexure test featuring uniform pressure loading, were employed in this study for purposes of evaluating the effects of size and stress state, respectively, on strength. Each test was specially designed to minimize spurious stresses and was qualified by experimentally calibrating and verifying stress analyses used to calculate strengths. The special care taken in the qualification of the strength tests was necessary to eliminate sources of strength variations other than those arising from the variability of strength-controlling flaws.

Both the pressure-on-disk and the uniaxial bend tests were conducted in a universal testing machine (Model TT-D, Instron Corporation, Canton, Mass.). The test environment and the stressing rates were selected so that slow-crack-growth effects on strength were either minimized (dry nitrogen gas, 250 MPa/s) or promoted (water, 2.5 MPa/s).

Pressure-on-Disk Test—A self-contained hydraulic test cell, used for uniform pressure loading of disk specimens, is shown in the schematic of Fig. 1. The details of the design, test qualification and procedures, and the advantages of this test over conventional bend tests were discussed in our earlier paper [10]. Only those test details needed for calculation of fracture stresses and application of fracture statistics are presented here. The disk specimen, supported along a concentric line support near its periphery, is subjected to axisymmetric bending by the lateral hydraulic pressure. The radial, σ_r, and tangential, σ_t, stresses are maximum and equal at the disk center and decrease along a radius according to the following relations obtained from simple plate theory [10,12]

$$\sigma_r = \sigma_u \left[1 - \alpha \left(\frac{r}{r_1} \right)^2 \right] \tag{1}$$

$$\sigma_t = \sigma_u \left[1 - \beta \left(\frac{r}{r_1} \right)^2 \right] \tag{2}$$

where σ_u is the maximum balanced stress at the disk center

$$\sigma_u = \frac{3pr_1^2}{8t^2} \left\{ 2(1 - \nu) + (1 + 3\nu)\left(\frac{r_2}{r_1} \right)^2 - 4(1 + \nu)\left(\frac{r_2}{r_1} \right) \ln \left(\frac{r_2}{r_1} \right) \right\}$$
$$+ \frac{(3 + \nu)p}{4(1 - \nu)} \tag{3}$$

and p is the uniform applied pressure, r_1 the support radius, t the disk specimen thickness, r_2 the disk specimen radius, and ν the Poisson's ratio for the specimen material; α and β are two nondimensional parameters that define the stress gradients along a radius

$$\alpha = \frac{3p(3 + \nu)r_1^2}{8\sigma_u t^2} \tag{4}$$

and

$$\beta = \frac{3p(1 + 3\nu)r_1^2}{8\sigma_u t^2} \tag{5}$$

Experimental stress analysis by strain measurements has confirmed the validity of the plate theory solutions for disks of nominal dimensions $t = 2.5$ mm, $r_2 = 16$ mm, and a support radius, $r_1 = 15$ mm [10]. The agreement was within 3% at all locations; this maximum discrepancy, which occurred at the disk center, was due to the limitations of the simple plate theory because it neglects the contributions of shear stresses to the deflection of the plate [12].

The maximum bending stress in alumina disk specimens was also calibrated by strain-gage measurements and by numerical analyses, employing an axisymmetric finite element model [13]. The experimental results, along with the

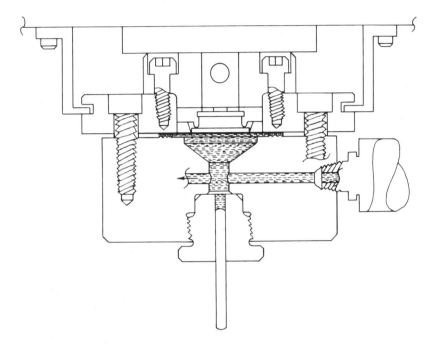

FIG. 1—*Hydraulic test cell designed for uniform pressure loading of disk specimens. (For details see Ref 1.)*

plate theory and finite element solutions obtained, are compared in Fig. 2. The strain measurements agreed with the finite element solutions to within 1%, and they were about 3% greater than the plate theory solution. The strain-gage calibration was used to calculate the fracture stresses, σ_u, for the alumina disks from fracture pressures, p. The values of the elastic modulus, $E = 316.3$ GPa, and the Poisson's ratio, $\nu = 0.22$, for the alumina used in this calibration were measured separately in direct compression experiments.

Four-Point Bend Test—The test fixtures used in the four-point bend tests have been described by Hoagland, Marschall, and Duckworth [14]. The design of these fixtures was based on an analysis that evaluated and minimized errors from specific sources in bend testing of brittle materials. These included errors from unequal moments, twisting, wedging, and friction. Figure 2 also compares the strains measured on an alumina bend specimen of nominal dimensions 2.5 by 5 by 38 mm subjected to four-point bending with a support span of 32 mm and a loading span of 19 mm, with the prediction of the conventional bending formula. The agreement is within 2%. As in the case of the disk specimens, strain-gage calibration was used to convert fracture loads to fracture stresses for all the four-point bend specimens of alumina.

Three-Point Bend Test—The four-point bend test fixtures were also used for three-point bend tests with a slight modification of the loading fixtures. It

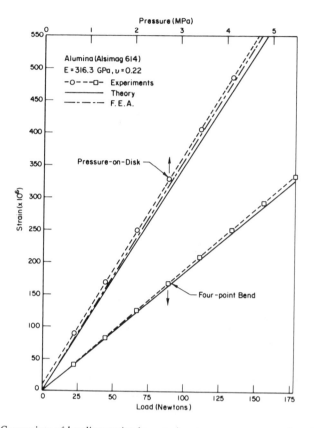

FIG. 2—*Comparison of bending strains from strain-gage measurements and strains predicted from theories for pressure-on-disk and four-point bend specimens of alumina.*

was not possible to calibrate directly the bending stresses in three-point loading because of the stress gradients. The test fixtures used, however, were of a design that minimized all errors except that due to wedging at the center loading line. The error from wedging stresses is a function of the thickness-to-support-span ratio. For the test geometry employed, the magnitude of this error is 0.7 or 1.5%, depending on the specific details of the wedging problem [14,15]. No attempt was made to incorporate this correction to the bending stress, and the three-point bend strength values reported in this paper are based on the conventional bending formula.

Test Material and Surface Finish

Test Material—The alumina ceramic (Alsimag 614, Technical Ceramic Products Div., 3M Co., Chattanooga, Tenn.) used in the present study is identical to and from the same lot as those used in two earlier studies [10,16].

The ceramic has a nominal 96% alumina (Al_2O_3) composition, with an average grain size of 5 μm. The fracture toughness of the ceramic, K_{Ic}, as determined by the double torsion test, was 3.84 $\pm$ 0.05 MPa m$^{1/2}$ [16]. Subcritical crack growth in the ceramic in water follows the empirical power law relation

$$V = V_0 \left(\frac{K_I}{K_{Ic}} \right)^n \qquad (6)$$

where V is the crack growth rate and K_I is the applied stress intensity factor. Values measured for the empirical crack growth parameters, V_0 and n, were 3.5 m/s and 42, respectively [16].

Surface Finish—The surface finish on the disk and bend specimens placed in tension required special attention for several reasons. First, the ceramic failed from surface flaws in all three of the tests in the two testing environments. Thus, a reproducible and consistent surface finish was required so that the same flaw size distribution determined the strengths in the different specimen and test geometries. Second, the selected flaw size distribution should be isotropic; that is, the strength-controlling flaws should not be oriented in preferred directions. The latter requirement derives from the fact that in statistical fracture theories one makes the assumption that all flaw orientations with respect to the principal stress directions are equally likely. A diamond wheel grinding finish was desirable to obtain this consistent and reproducible finish. But conventional grit sizes used in grinding, for example, 240 and 320 grit, produced unacceptable levels of anisotropy in strengths. This anisotropy in strengths was assessed by comparing the fracture strengths of four-point bend bars ground parallel to the long axis (longitudinal) with the strengths of test bars ground perpendicular to the long axis (transverse). For a 320-grit surface finish the anisotropy was 8%—that is, the mean longitudinal strength was 8% greater than the mean transverse strength. For a 600-grit finish, however, the strength anisotropy decreased substantially to 3%, and, further, the coefficients of variation of strength for the two orientations were very similar—approximately 5%. These results were taken to indicate nearly random and isotropic flaw size distributions resulting from the 600-grit grinding surface finish. This interpretation was supported by the fracture patterns exhibited by the disk specimens. A typical pattern in an alumina specimen is shown in Fig. 3. The fracture origins were located randomly near the disk center, and the initial fracture paths forming a "jog" were randomly oriented toward the grinding direction indicated by the top arrow in the figure.

All specimen surfaces placed in tension were finished with the 600-grit diamond wheel by removing the final 127 μm of the surface with a 1.27-mm cross-feed and a 5-μm downfeed per pass. Corners on the bend bars were rounded and polished with 1-μm diamond paste. Fracture origins in the bend bars were also located at random points on the tension surface.

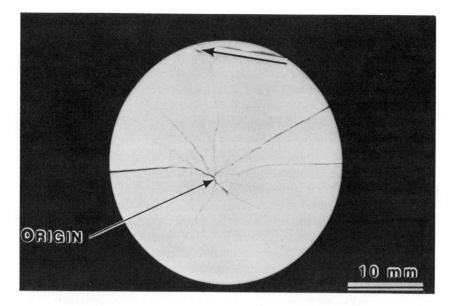

FIG. 3—*Fracture pattern in a pressure-on-disk test specimen of alumina. The arrow at the top indicates the direction of grinding during surface finishing.*

Statistical Analysis

In the analysis of the preliminary biaxial strength data obtained on alumina ceramic, Weibull's statistical theory of strength was applied to examine both size and stress state effects on strength [10]. As indicated earlier, Weibull's theory was adequate to account for the size effect, but it underestimated the biaxial strength. The reason for this discrepancy between the Weibull theory predictions and the measured strength values is not altogether clear. Weibull's theory for multiaxial stress states has generally been criticized on two accounts [17]: (1) the basis of extension of fracture statistics in the Weibull theory from the well-known uniaxial case to general multiaxial stress states is somewhat arbitrary; and (2) Weibull's theory implicitly assumes a normal stress fracture criterion. But experiments with controlled surface flaws in ceramics indicate that the normal stress fracture criterion is inadequate to treat fracture from cracks oriented at arbitrary angles to the principal stresses [18–20]. Batdorf and Heinisch [4], Evans [21], and Jayatilaka and Trustrum [22] have recently developed statistical fracture theories that overcome these limitations and incorporate in their theoretical framework fracture criteria derived from experiments or fracture mechanics theory. We employed Batdorf's analysis because it leads to closed-form solutions in the case of the pressure-on-disk test and also reduces to the commonly used Weibull theory when the normal stress fracture criterion is invoked [23].

Formulation of Fracture Statistics

According to Batdorf's theory [4] the probability of fracture, F, of a brittle solid, subjected to a general state of stress and failing from a distribution of surface cracks, is given by

$$F = 1 - \exp\left[-\int_A \int_0^\sigma \frac{\Omega dN(\sigma_c)}{4\pi d\sigma_c} d\sigma_c dA \right] \tag{7}$$

where σ_c is a critical stress, defined as the remote tensile stress that will cause fracture when applied normal to the crack plane; $N(\sigma_c)$ is a function representing the number of surface cracks per unit area that will cause fracture when tension normal to the crack plane is $\leq \sigma_c$; Ω is a solid angle in the principal stress space enclosing the normals to crack planes so that an effective stress, σ_e, which is a function of the principal stresses and crack orientation, exceeds σ_c. In essence, $N(\sigma_c)$ is equivalent to a crack size distribution function and is a characteristic of the material surface condition. On the other hand, Ω is a function of both the applied stress state and the specific fracture criterion employed. In the special case of equitriaxial stress state (hydrostatic tension), $\Omega = 4\pi$, and then $N(\sigma_c)$ also defines the fracture stress distribution [4].

For analytical simplicity we assumed in this study that crack size distribution characterizing the surface flaws from grinding is of the form

$$N(\sigma_c) = \bar{k}(\sigma_c)^m \tag{8}$$

where $\bar{k}$ and m are location and shape parameters, respectively, in analogy with the Weibull parameters. It will be shown later in the section on experimental results that this simple form of the flaw size distribution function is adequate to describe fracture strength distributions of alumina ceramic in the majority of cases.

Fracture Criteria for Brittle Materials

Griffith's fracture criterion, based on an energy-balance argument, is the most recognized and accepted fracture criterion for brittle materials when the plane of the dominant crack is normal to the applied tensile stress [24,25]. But, for an arbitrary crack orientation toward the applied stresses, which is the more likely situation in structural components, both remote tensile stress normal to the crack plane and the in-plane shear stress parallel to the crack plane influence the fracture condition. Several fracture criteria based on fracture mechanics theory have been advanced to treat this mixed-mode crack extension problem, and they have been reviewed by Swedlow [26] and Duckworth and Rosenfield [1]. In general, all the different fracture criteria that predict noncoplanar crack extension are very similar in their predictions [26], and the following empirical equation is a good fit to all of them [27]

$$\frac{\sigma_N}{\sigma_c} + \frac{3}{2} \left(\frac{\tau}{\sigma_c} \right)^2 = 1 \tag{9}$$

where τ is the remote shear parallel to the crack plane. An alternate fracture criterion that has often been used to treat mixed-mode fracture in ceramics [4,21] is a strain-energy-release rate criterion that assumes coplanar crack extension [28]

$$\frac{\sigma_N^2}{\sigma_c^2} + \frac{\tau^2}{\sigma_c^2} = 1 \tag{10}$$

Comparison of the data available on ceramics [18–20] with these mixed-mode fracture criteria indicates that one cannot make a clear discrimination between the two shear-sensitive fracture criteria. We employed the coplanar strain-energy-release rate criterion (Eq 10), because it was simpler than the noncoplanar criterion (Eq 9). For purposes of comparison, we also present fracture statistics formulations that employ the normal stress fracture criterion, $\sigma_N = \sigma_c$, to gain insight into the relative influence of fracture criteria on predicted fracture probabilities.

Fracture Statistics

Equations for cumulative fracture probabilities for specimens used in the three strength tests were formulated by first defining Ω for the selected fracture criterion and the stress state in the test specimens and then applying Eq 7 along with the crack size distribution function, $N(\sigma_c)$ (Eq 8). As shown by Batdorf and Crose [29], closed-form expressions can be derived for Ω for analytically simple fracture criteria in uniaxial and balanced biaxial tension stress states.

Normal Stress Fracture Criterion—For the normal stress fracture criterion and the uniaxial stress state present in the four-point and the three-point bend specimens, it has been shown that [29]

$$\Omega = 4\pi \left[1 - \left(\frac{\sigma_c}{\sigma} \right)^{1/2} \right] \tag{11}$$

With the introduction of the variation of σ with position, substitution of Eqs 11 and 8 in Eq 7, and integration, the following final expressions for the probability of fracture for the four-point and three-point bend specimens are produced:

Four-point bend

$$F = 1 - \exp \left[-\frac{\bar{k}\sigma_f^m}{(2m+1)} \left\{ \frac{bL(ma/L+1)(m+1) + tL(ma/L+1)}{(m+1)^2} \right\} \right] \tag{12}$$

Three-point bend

$$F = 1 - \exp\left[-\frac{\bar{k}\sigma_t^m}{(2m + 1)} \left\{ \frac{bL(m + 1) + Lt}{(m + 1)^2} \right\} \right] \qquad (13)$$

where σ_f and σ_t are the maximum tensile stresses in the four-point and the three-point bend specimens, and b, t, L, and a are specimen thickness, width, support span, and loading span (in four-point bend tests), respectively.

For a balanced biaxial stress state, the normal stress fracture criterion leads to

$$\Omega = 4\pi\left[1 - \left(\frac{\sigma_c}{\sigma} \right) \right]^{1/2} \qquad (14)$$

In the pressure-on-disk test, the stress state is precisely balanced biaxial only at the disk center; it deviates from this stress state with increasing radial position because the radial stress decreases slightly more rapidly ($\alpha \sim 1$) than does the tangential stress ($\beta \sim 0.52$). We can, however, obtain lower and upper-bound failure probabilities for the disk specimens by assuming balanced biaxial stresses equal to the radial stress (Eq 1) and tangential stress (Eq 2), respectively. Following this procedure, substitution of Eqs 8 and 14 in Eq 7 and integration leads to the following relation for the lower-bound failure probability for a disk specimen subjected to a center maximum stress, σ_u:

Pressure-on-disk

$$F = 1 - \exp\left[-\frac{\bar{k}\sigma_u^m}{(2m + 1)} \left\{ \frac{\sqrt{\pi}\,m\Gamma(m)\pi r_1^2}{(m + 1)\Gamma(m + 1/2)\alpha} \right\} \right] \qquad (15)$$

The upper-bound failure probability for the disk specimen is given by an identical equation in which the stress-gradient factor β replaces α.

Coplanar Strain-Energy-Release Rate Criterion

For the uniaxial stress state and coplanar strain-energy-release rate criterion (Eq 10), Ω is derived by a method analogous to that used for the normal stress criterion [4]

$$\Omega = 4\pi\left[1 - \left(\frac{\sigma_c}{\sigma} \right) \right] \qquad (16)$$

and the fracture probabilities for the three-point and four-point bend specimens are given by the following relations:

Four-point bend

$$F = 1 - \exp\left[-\frac{\bar{k}\sigma_f^m}{(m+1)} \left\{ \frac{bL(ma/L+1)(m+1) + tL(ma/L+1)}{(m+1)^2} \right\} \right]$$

(17)

Three-point bend

$$F = 1 - \exp\left[-\frac{\bar{k}\sigma_t^m}{(m+1)} \left\{ \frac{bL(m+1) + Lt}{(m+1)^2} \right\} \right]$$ (18)

Comparison of Eqs 17 and 18 with Eqs 12 and 13 reveals that for the same flaw distribution (same $\bar{k}$ and m), the coplanar strain-energy-release rate criterion predicts a higher fracture probability, F, for the same nominal applied stresses in the bend specimens than does the normal stress failure criterion. This is to be expected since the normal stress criterion does not account for the influence of the remote shear stress on fracture. A second interesting observation is that the ratio of the fracture stresses, σ_f and σ_t, at the same failure probability is independent of the failure criterion; that is, the size effect in uniaxial stressing is not dependent on the fracture criterion in the present formulation. Thus, Batdorf's generalized statistical theory [4] predicts the same size effect on strength as does the classic Weibull theory [5].

For a balanced biaxial stress state the strain-energy-release rate criterion obtains

$$\Omega = 4\pi \left[1 - \left(\frac{\sigma_c}{\sigma}\right)^2 \right]^{1/2}$$ (19)

Following the procedure used in the case of the normal stress fracture criterion, one can obtain the following relation for a lower-bound failure probability for a disk specimen subjected to a center maximum stress, σ_u:

Pressure-on-disk

$$F = 1 - \exp\left[-\frac{\bar{k}\sigma_u^m}{(m+1)} \left\{ \frac{\sqrt{\pi}m\Gamma(m/2)\pi r_1^2}{2(m+1)\Gamma[(m+1)/2]\alpha} \right\} \right]$$ (20)

and, as before, upper-bound failure probability is given by an identical equation with β replacing α.

As in the case of uniaxial bend tests, Eq 20 based on the strain-energy-release rate criterion predicts a higher failure probability than does the normal stress failure criterion (Eq 15). A more significant result, however, is that the stress state effect on strength, which is given by the ratio of fracture stresses in biaxial and uniaxial tests, (σ_u/σ_f) or (σ_u/σ_t) (at the same fracture probability), does depend on the selected fracture criterion. This result provided the motivation for examining the strength data in terms of the generalized statistical fracture theory rather than the classic Weibull theory. A critical analysis of the strength data obtained in the three tests, in terms of the fracture probability equations derived in this section, is presented in the following section.

Experimental Results

In the following subsections, the fracture strength data obtained in the different tests and the environmental and testing conditions are discussed with reference to Weibull plots, that is, plots of $\ell n \; \ell n [1/(1 - F)]$ versus $\ell n \; \sigma$, where F, the fracture probability, was defined by the relation [30]

$$F = \frac{(i - 0.5)}{S} \qquad (21)$$

where i is a ranking number of a specimen in a sample size S in increasing order of fracture stress, σ. Figures 4 and 5 give these plots.

In the case of the uniaxial bend tests, the samples were made up of equal numbers of longitudinal (filled symbols) and transverse (open symbols) ground specimens. This was done as an additional attempt to learn whether flaw-orientation anisotropy might exist in directionally ground specimens. The relative positions of the filled and the unfilled symbols in each strength distribution indicate the strength isotropy.

In each comparison of experimentally obtained and statistically predicted strengths, the four-point bend strengths, σ_f, were used to establish reference strength distribution parameters, m and σ_0, where σ_0, a scale parameter, is related to the flaw size distribution parameters, $\overline{k}$ and m, by either $\sigma_0 = [(2m + 1)/\overline{k}]^{1/m}$ (maximum normal stress fracture theory—Eq 12) or $\sigma_0 = [(m + 1)/\overline{k}]^{1/m}$ (strain-energy-release rate theory—Eq 17), depending on the specific fracture criterion employed. Values of m and σ_0 were obtained from the four-point bend strength data and Eqs 12 or 17 by least-square linear regression. The linear-regression method rather than the maximum-likelihood method was used to evaluate m and σ_0, first, because the linear-regression method is simpler, and second, because for the sample sizes involved the two methods give nearly identical parameters, especially when fracture probability is defined by Eq 21 [30].

The values of m and σ_0 derived from the linear-regression fit to the four-point bend data (the heavy full line in Figs. 4 and 5) were used to predict the three-point and biaxial strengths using the fracture probability formulations derived in the previous section. The agreements obtained between the statistical theory predictions (full lines) and the experimentally measured fracture strengths (dotted lines) for the three-point and biaxial specimens indicated

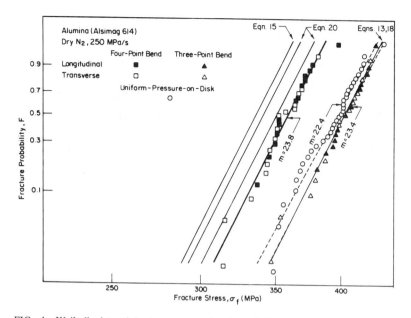

FIG. 4—*Weibull plots of fracture stresses for sintered alumina under slow-crack-growth-inhibiting conditions.*

the validity of the statistical fracture theories in predicting size and stress state effects on strengths of the alumina ceramic, respectively.

Fracture Strengths in an Inert Environment

Figure 4 summarizes the fracture strength data for the alumina ceramic tested in a dry nitrogen gas environment at a stressing rate of 250 MPa/s, conditions that inhibited slow crack growth. The linearity of the plots supported application of the two-parameter distribution function (Eq 8). The slopes of the plots, which give the shape parameter m, were nearly equal ($m \sim 23$) for the three tests. The longitudinally and transversely ground specimens exhibited nearly identical fracture stresses and stress distributions. These two observations indicated that a single flaw population was sampled in all three of the tests.

A shape parameter, $m = 23.8$, and a scale parameter, $\sigma_0 = 247$ MPa, obtained from the four-point bend data were used to predict the three-point bend strengths (from Eqs 13 or 18) and the biaxial strengths from Eqs 15 and 20. As seen in the figure, excellent agreement was obtained between the measured and the predicted three-point bend strengths, thus showing that the generalized statistical fracture theory is completely adequate to account for size effects on the strength of alumina ceramic. A similar conclusion was reached in our earlier analysis with the Weibull theory [10]. It is again empha-

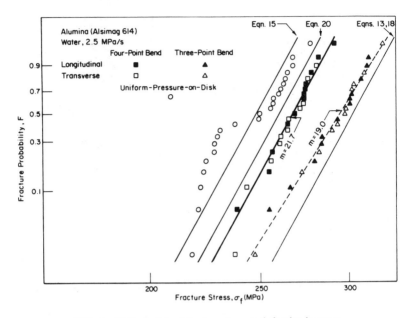

FIG. 5—*Weibull plots of fracture stresses of alumina in water.*

sized here that both the normal stress fracture criterion and the strain-energy-release rate fracture criterion predict the same size effect in the present analysis, since the intrinsic differences arising from the two criteria are normalized in the definition of the scale parameter, σ_0.

The intriguing result in Fig. 4 is the relatively large measured biaxial strength of the alumina ceramic. The indicated theoretical predictions are those based on the strain-energy-release rate criterion, upper and lower-bound fracture probabilities (Eq 20), and upper-bound fracture probability predicted by the normal stress fracture criterion (Eq 15). The predicted median biaxial strengths are in the range of 335 to 350 MPa, and the observed median is 400 MPa. This 15 to 20% difference between theory and experiment must be considered significant in view of the small ($\sim 2\%$) uncertainties associated with fracture stress measurements; statistical fracture theories in their present form, therefore, are inadequate. They fail to predict this observed biaxial strengthening. Figure 4 shows that the normal stress fracture criterion predicts lower values of biaxial strength than the strain-energy-release rate criterion, but the relative difference between the predictions of the two criteria is small. This latter difference is a function of the parameter, m, and it becomes large and significant as m decreases. Thus, the selection of the appropriate fracture criterion is of greater importance when dealing with materials with lower m values.

The discrepancy noted here for the alumina ceramic between the measured biaxial strengths and the strengths calculated from the statistical theories is similar to our finding in the earlier study on glass ceramic [6]. That study also indicated that tests in water did not show a similar discrepancy. This earlier observation provided a motivation to examine whether there was a similar correlation between biaxial strengthening and the extent of slow crack growth in alumina ceramic.

Fracture Strengths in Water

Weibull plots of fracture stresses for alumina specimens tested in water at low stressing rates ($\dot{\sigma} \sim 2.5$ MPa/s) are shown in Fig. 5. Several significant results can be seen in this figure. As was the case in dry nitrogen gas tests, the measured three-point bend strengths are in accord with the statistical predictions based on parameters from four-point bend strengths, $m = 21.7$ and $\sigma_0 = 178$ MPa, but the agreement is not as close as for the dry nitrogen gas tests (Fig. 4). A more important result in Fig. 5, however, is the biaxial strengths in relation to the statistical theory predictions. Unlike the results of the dry nitrogen gas tests, the biaxial strengths obtained in water are reasonably close to the theoretical predictions, although the biaxial strength distribution is quite nonlinear and tends to exhibit a lower-bound strength. In the high fracture stress regimen, the measured biaxial strengths are approximately in the middle of the two limiting bounds of fracture probability predicted by Eqs 15 and 20.

Thus, it appears that the subcritical crack growth associated with the water tests affects flaw morphology in such a way that both specimen size and stress state effects can be predicted with reasonable accuracy using available statistical fracture theories. In contrast, statistical theories grossly underestimate the biaxial strengths in inert environments, and there is a transition from an apparent biaxial strengthening to an expected biaxial weakening with the extent of slow crack growth, that is, with the size of the critical crack. Plausible mechanisms that might explain the biaxial strengthening will be considered in the discussion section.

The observed transition from biaxial strengthening (Fig. 4) to biaxial weakening (Fig. 5) indicates that the alumina ceramic is more susceptible to loss of strength through slow crack growth in a biaxial stress field. These differences in the relative susceptibilities to slow crack growth and strength degradation in the two stress states can be put in perspective by applying the well-known fracture mechanics relation for fracture strengths in dynamic fatigue conditions [31,32]

$$\sigma = \left[\frac{2(n + 1)\dot{\sigma} K_{Ic}^2}{(n - 2) V_0 Y^2} \right]^{1/(n + 1)} \sigma_I^{(n - 2)/(n + 1)} \tag{22}$$

where σ is the fracture strength obtained in a test employing a constant stressing rate, $\dot{\sigma}$; σ_I is the inert strength; and Y is a fracture mechanics parameter in

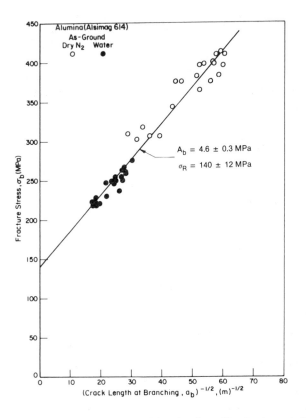

FIG. 6—*Relationship of fracture stress to crack length at branching for sintered alumina disks.*

the stress intensity factor relation that combines the influences of crack shape, free surface, and stress gradient effects on the stress intensity factor. For small semicircular surface cracks, $Y \sim 1.24$ [*33*]. Equation 22 was used iteratively to derive the effective slow-crack-growth exponents, n, for correctly predicting strengths in water from the inert strengths in nitrogen gas. Values for the parameters $V_0 = 3.5$ m/s and $K_{Ic} = 3.84$ MPa m$^{1/2}$, were obtained in double torsion tests [*16*] for use in these calculations.

As expected, the biaxial strength data gave a low value of $n = 33$, while the corresponding values for the four-point and three-point bend strengths were 50 and 48, respectively. It is interesting to note that the uniaxial n values are greater than that measured in a double torsion test ($n = 42$), whereas the biaxial value is significantly lower. This result has a significant bearing on reliability assessments of brittle structural components for long-term applications; it implies that the current practice of using n values established in uniaxial bend tests or fracture mechanics tests may not be conservative for components subjected to general multiaxial loading.

The biaxial strength distribution in the water tests (Fig. 5) was unique in that it exhibited three stages. At the high fracture stresses the strength distribution was similar to the uniaxial strength distributions. There was a transition at about the median strength level, and below this there was a high slope. This high-slope, low-fracture-stress regime suggested the possible existence of a lower-bound strength. Such a lower bound is usually found in specimens after they are proof tested [34]. An apparent lower-bound strength would also be expected if the specimen surfaces have a compressive residual stress. Green and Lange [35] have measured compressive residual stresses on ground surfaces of polycrystalline alumina by X-ray line broadening analysis. The presence of similar compressive residual stresses in our specimens was suggested by indirect evidence. The fracture morphology of the disk specimens (see Fig. 3) changed systematically with the fracture stress level. In particular, the length of the crack from the origin to the point of branching, a_b, increased with decreasing fracture stress at the branching point, σ_b, as shown in Fig. 6. The experimentally observed relation was of the form

$$\sigma_b = \sigma_R + A_b a_b^{-1/2} \tag{23}$$

where A_b and σ_R are empirical constants. Although the functional form of the crack branching relation (Eq 23) is similar to those obtained in many other crack branching studies [36], the large intercept on the fracture stress axis is not commonly observed. Such a fracture stress intercept can be expected if the test specimen is under compressive residual stress. The crack driving force (or branching stress intensity factor) would then be expected to be proportional to the net tensile stress, $(\sigma_b - \sigma_R)$. If the uniaxial strength testing were extended to provide data which define a low-stress, low-probability regime, by testing more or larger specimens, the transition shown by the biaxial data could be expected in the uniaxial plots, if it is the result of a residual surface stress.

Discussion

The objective of this study was to evaluate fracture strengths of an alumina ceramic in three qualified tests and then critically examine the relative strengths in terms of statistical fracture theory. A number of significant results emerged from this study. The specimen size effect on strength, as derived from the four-point and three-point bend tests, is in good agreement with theoretical predictions in the case of both inert strengths and strengths affected by slow crack growth. Statistical fracture theory, however, accounted for the stress state effect on strength only in the case of tests in water. Inert biaxial strengths were significantly greater than predicted by statistical theory, irrespective of the fracture criterion employed.

One explanation for the experimental observation of biaxial strengthening in tension-tension stress states was suggested by Babel and Sines [2]. These investigators developed a generalized Griffith theory to treat fracture from bulk defects such as pores by modeling them as two-dimensional elliptical stress concentrators. For small aspect ratios of the ellipse (representing bulk defects such as pores), the maximum stress concentration at the defect is smaller under biaxial tension than under uniaxial tension. A critical maximum stress fracture criterion, therefore, predicts higher fracture stress under biaxial tension than under uniaxial tension. In the other extreme of large aspect ratio of the ellipse (representing crack like flaws), the maximum stress concentration is unaffected by the applied stress state. Thus, the theory of Babel and Sines [2] suggests that ceramics that fail from bulk defects such as pores tend to exhibit biaxial strengthening, while dense ceramics that fail from crack like flaws would exhibit biaxial weakening that is normally expected from statistical considerations. A fracture mechanics analog of the Babel–Sines theory was developed by the present authors by modeling the flaws as spherical pores with annular cracks [37]. Remote fracture stresses were calculated for different stress states by employing a critical stress intensity fracture criterion. The results of this refined theory differed from the Babel–Sines theory in two respects. The fracture stresses for the pore-crack complex were sensitive to the applied stress state only when the annular cracks were comparable to or smaller than the pores. Second, the magnitude of the biaxial strengthening predicted was smaller than that of the Babel–Sines theory.

Fracture surfaces of selected alumina specimens tested in this study were examined to see if the previous explanation might be applicable. Figure 7a and b shows the fracture surface in the vicinity of the fracture origin in a biaxial disk specimen of alumina tested in dry nitrogen gas. The fracture surface does show pores comparable in size to the grains. It is difficult to assess their influence on fractures, however, because of the general difficulty in determining critical flaw sizes by fractography. The basic concept of the previous theories, however, is still appealing because it is consistent with the observed transition from biaxial strengthening for small flaws to the biaxial weakening observed and expected for large cracklike flaws.

The second significant result obtained in this study is the enhanced degradation of strength (that is, the lower effective slow-crack-growth exponent, n) observed in the biaxial tests. Pletka and Wiederhorn [7] have also reported a similar trend in an alumina ceramic. Although some qualitative ideas based on crack-microstructure interactions have been put forth for this apparent enhancement of slow crack growth in biaxial tension stress fields [38], the phenomenon is still largely unexplained and warrants further work. But it is important to emphasize here that the biaxial strength data available to date strongly suggest a reduced resistance to crack growth in ceramics in biaxial stress fields. This implies that slow-crack-growth exponents derived from fracture mechanics or uniaxial strength tests alone cannot be relied upon to obtain conservative life predictions for components under general loading.

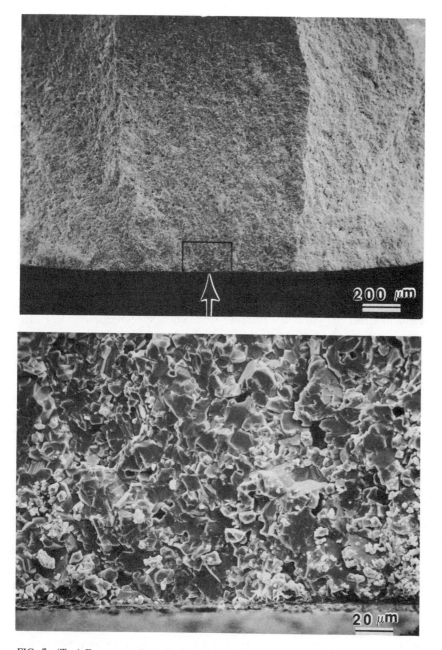

FIG. 7—(Top) *Fracture surface of a biaxial disk specimen.* (Bottom) *Fracture surface in the vicinity of the origin.*

Conclusions

1. Fracture strengths of the alumina ceramic evaluated in both inert and water environments in three-point and four-point bend tests closely followed the size effect predictions of statistical fracture theory.

2. Biaxial fracture strengths of the alumina were in the predicted range for tests in water, but in the inert environment the measured strengths were significantly greater than the theoretical predictions, irrespective of the fracture criteria employed in the theory.

3. The observed transition of the relative strength values from an apparent biaxial strengthening to a biaxial weakening implied enhanced strength degradation by means of slow crack growth in biaxial tension.

4. Evidence was obtained of the presence of a residual compressive stress on the ground surfaces of the specimens, which could explain an observed indication in Weibull plots of a lower-bound strength.

Acknowledgments

The authors are grateful to Dr. C. Tsay for his contributions on the finite element stress analysis of the pressure-loaded disk specimen and to P. R. Held for his assistance on experimental work. This research was supported by the Division of Materials Sciences, Office of Energy Research, U.S. Department of Energy, under Contract W-7504-Eng-92, Task 127.

References

[1] Duckworth, W. H. and Rosenfield, A. R., "Effects of Stress State on Ceramic Strength," in *Energy and Ceramics*, F. Vincenzini, Ed., Elsevier, Amsterdam, 1980, pp. 645-660.

[2] Babel, H. W. and Sines, G., "A Biaxial Fracture Criterion for Porous Brittle Materials," *Journal of Basic Engineering*, Vol. 90, 1968, pp. 285-291.

[3] Giovan, M. N. and Sines, G., "Biaxial and Uniaxial Data for Statistical Comparisons of a Ceramic's Strength," *Journal of the American Ceramic Society*, Vol. 62, Nos. 9-10, 1979, pp. 510-515.

[4] Batdorf, S. B. and Heinisch, H. L., Jr., "Weakest Link Theory Reformulated for Arbitrary Fracture Criterion," *Journal of the American Ceramic Society*, Vol. 61, Nos. 7-8, 1978, pp. 355-358.

[5] Weibull, W., "A Statistical Theory of the Strength of Materials," *Ingeniors vetenskapsakademiens*, Handlingar Nr. 151, 1939.

[6] Shetty, D. K., Rosenfield, A. R., Bansal, G. K., and Duckworth, W. H., "Biaxial Fracture Studies of a Glass-Ceramic," *Journal of the American Ceramic Society*, Vol. 64, No. 1, 1981, pp. 1-4.

[7] Pletka, B. J. and Wiederhorn, S. M., "A Comparison of Failure Predictions by Strength and Fracture Mechanics Techniques," *Journal of Material Science*, Vol. 17, 1982, pp. 1247-1268.

[8] Petrovic, J. J. and Stout, M. G., "Fracture of A_2O_3 in Combined Tension/Torsion: I, Experiments," *Journal of the American Ceramic Society*, Vol. 64, No. 11, 1981, pp. 656-660.

[9] Petrovic, J. J. and Stout, M. G., "Fracture of A_2O_3 in Combined Tension/Torsion: II, Weibull Theory," *Journal of the American Ceramic Society*, Vol. 64, No. 11, 1981, pp. 661-665.

[10] Shetty, D. K., Rosenfield, A. R., Duckworth, W. H., and Held, P. R., "A Biaxial-Flexure Test for Evaluating Ceramic Strengths," *Journal of the American Ceramic Society*, Vol. 66, No. 1, 1983, pp. 36-42.

[11] Batdorf, S. B., "Some Approximate Treatments of Fracture Statistics for Polyaxial Tension," *International Journal of Fracture*, Vol. 13, No. 1, 1977, pp. 5-11.

[12] Szilard, R., in *Theory and Analyses of Plates, Classical and Numerical Methods*, Prentice-Hall, Englewood Cliffs, N. J., 1974, p. 638.

[13] *MSC/NASTRAN User and Application Manuals*, The MacNeal-Schwendler Corporation, Los Angeles, Calif., Revised, May 1980.

[14] Hoagland, R. G., Marschall, C. W., and Duckworth, W. H., "Reduction of Errors in Ceramic Bend Tests," *Journal of the American Ceramic Society*, Vol. 59, No. 5-6, 1976, pp. 189-192.

[15] Baratta, F. I., "Requirements for Flexure Testing of Brittle Materials," Technical Report AMMRC TR 82-80, Army Materials and Mechanics Research Center, Watertown, Mass., April 1982.

[16] Bansal, G. G., Duckworth, W. H., and Niesz, D. E., "Strength-Size Relations in Ceramic Materials: Investigation of an Alumina Ceramic," *Journal of the American Ceramic Society*, Vol. 59, Nos. 11-12, 1976, pp. 472-478.

[17] Batdorf, S. B., "New Light on Weibull Theory," *Nuclear Engineering and Design*, Vol. 47, 1978, pp. 267-272.

[18] Petrovic, J. J. and Mendiratta, M. G., "Mixed-Mode Fracture from Controlled Surface Flaws in Hot-Pressed Si_3N_4," *Journal of the American Ceramic Society*, Vol. 59, Nos. 3-4, 1976, pp. 163-167.

[19] Freiman, S. W., Gonzalez, A. C., and Mecholsky, J. J., "Mixed-Mode Fracture in Soda-Lime Glass," *Journal of the American Ceramic Society*, Vol. 62, Nos. 3-4, 1979, pp. 206-208.

[20] Petrovic, J. J., "Controlled Surface Flaw Fracture in Tension and Torsion," *Fracture Mechanics of Ceramics*, Vol. 5, R. C. Bradt, D. P. H. Hasselman, F. F. Lange, and A. G. Evans, Eds., Plenum, New York, 1983, pp. 63-75.

[21] Evans, A. G., "A General Approach for the Statistical Analysis of Multiaxial Fracture," *Journal of the American Ceramic Society*, Vol. 61, Nos. 7-8, 1978, pp. 303-308.

[22] Jayatilaka, A. De S. and Trustrum, K., "Statistical Approach to Brittle Fracture," *Journal of Materials Science*, Vol. 12, 1977, pp. 1426-1430.

[23] Batdorf, S. B., "Weibull Statistics for Polyaxial Stress States," *Journal of the American Ceramic Society*, Vol. 57, No. 1, 1974, pp. 44-45.

[24] Griffith, A. A., "The Phenomena of Rupture and Flow in Solids," *Philosophical Transactions*, Royal Society, Vol. 221A, 1920, pp. 163-198.

[25] Griffith, A. A., "The Theory of Rupture," *Proceedings*, First International Congress on Applied Mechanics, Delft, Netherlands, 1924, pp. 55-63.

[26] Swedlow, J. L., "Criteria for Growth of the Angled Crack," in *Cracks and Fracture, ASTM STP 601*, American Society for Testing and Materials, Philadelphia, 1976, pp. 506-521.

[27] Palaniswamy, K. and Kanuss, W. G., "On the Problem of Crack Extension in Brittle Solids Under General Loading," *Mechanics Today*, Vol. 4, S. Nemat-Nasser, Ed., Pergamon, Oxford, 1978, pp. 87-148.

[28] Paris, P. C. and Sih, G. C., "Stress Analysis of Cracks," in *Fracture Toughness Testing and Its Applications, ASTM STP 560*, American Society for Testing and Materials, Philadelphia, 1965, pp. 30-81.

[29] Batdorf, S. B. and Crose, J. G., "A Statistical Theory for the Fracture of Brittle Structures Subjected to Polyaxial Stress States," *Journal of Applied Mechanics*, Vol. 41, No. 2, 1974, pp. 459-464.

[30] Trustrum, K. and Jayatilaka, A. De S., "On Estimating the Weibull Modulus for a Brittle Material," *Journal of Material Science*, Vol. 14, 1979, pp. 1080-1084.

[31] Evans, A. G., "Slow Crack Growth in Brittle Materials Under Dynamic Loading Conditions," *International Journal of Fracture Mechanics*, Vol. 10, No. 2, 1974, pp. 251-259.

[32] Wiederhorn, S. M., "Subcritical Crack Growth in Ceramics," in *Fracture Mechanics of Ceramics*, Vol. 2, R. C. Bradt, D. P. H. Hasselman, and F. F. Lange, Eds., Plenum, New York, 1974, pp. 613-646.

[33] Scott, P. M. and Thorpe, T. W., "A Critical Review of Crack Tip Stress Intensity Factors for Semi-elliptic Cracks," *Fatigue of Engineering Materials and Structures*, Vol. 4, No. 4, 1981, pp. 291-309.

[34] Fuller, E. R., Jr., Wiederhorn, S. M., Ritter, J. E., Jr., and Oates, P. B., "Proof Testing of Ceramics," *Journal of Material Science*, Vol. 15, 1980, pp. 2282-2295.

[35] Green, D. J. and Lange, F. F., "Measurement of Residual Surface Stresses in Al_2O_3/ZrO_2 Composites" (abstract), *American Ceramic Society Bulletin*, Vol. 61, No. 3, 1982, p. 338.

[36] Shetty, D. K., Rosenfield, A. R., and Duckworth, W. H., "Crack Branching in Ceramic Disks Subjected to Biaxial Flexure," *Journal of the American Ceramic Society*, Vol. 66, No. 1, 1983, pp. C10–C12.

[37] Shetty, D. K., Rosenfield, A. R., and Duckworth, W. H., "Biaxial Stress State Effects on Strengths of Ceramics Failing from Pores," *Fracture Mechanics of Ceramics*, Vol. 5, R. C. Bradt, D. P. H. Hasselman, F. F. Lange, and A. G. Evans, Eds., Plenum, New York, 1983, pp. 531–542.

[38] Pletka, B. J. and Wiederhorn, S. M., "Subcritical Crack Growth in Glass-Ceramics," in *Fracture Mechanics of Ceramics*, Vol. 4, R. C. Bradt, D. P. H. Hasselman, and F. F. Lange, Eds., Plenum, New York, 1978, pp. 745–759.

Matthew B. Magida,[1] *Katherine A. Forrest,*[1] *and Thomas M. Heslin*[1]

Dynamic and Static Fatigue of a Machinable Glass Ceramic

REFERENCE: Magida, M. B., Forrest, K. A., and Heslin, T. M., **"Dynamic and Static Fatigue of a Machinable Glass Ceramic,"** *Methods for Assessing the Structural Reliability of Brittle Materials, ASTM STP 844,* S. W. Freiman and C. M. Hudson, Eds., American Society for Testing and Materials, Philadelphia, 1984, pp. 81–94.

ABSTRACT: The dynamic and static fatigue behavior of a machinable glass ceramic was investigated to assess its susceptibility to stress corrosion–induced delayed failure. Fracture mechanics techniques were used to analyze the results so that lifetime predictions for components of this material could be made. The resistance to subcritical crack growth of this material was concluded to be only moderate and was found to be dependent on the size of its microstructure.

KEY WORDS: glass ceramic, fracture mechanics, dynamic fatigue, static fatigue, lifetime predictions, proof test, reliability, structural reliability, brittle materials

Toward the end of this decade, the National Aeronautics and Space Administration (NASA) plans to launch an energetic gamma ray experiment telescope (EGRET) as part of the Gamma Ray Observatory. Included in the design of this scientific instrument is a machinable glass ceramic material[2] for structural support beams comprising a spark chamber stack. These beams are designed for a minimum ten-year lifetime under applied load. Many ceramic materials are subject to delayed failure at stresses below ultimate strengths through stress corrosion induced subcritical crack growth [1,2]. Therefore, the susceptibility of this glass ceramic to delayed failure was investigated using dynamic fatigue testing and static fatigue time to failure measurements. These techniques yielded independent estimates of the crack growth parameters from which design diagrams were constructed relating time to failure to applied stress as a function of failure probability. Furthermore, the parameters

[1]Materials engineer and ceramics engineers, respectively, National Aeronautics and Space Administration Goddard Space Flight Center, Greenbelt, Md. 20771.
[2]Macor, Code No. 9658, Corning Glass Works, Corning, N.Y.

were used to determine a stress level for a proof test to establish a maximum initial flaw size which would preclude failure for the anticipated load/time conditions.

Theory

Modeling the failure of a subcritically loaded structure as the extension of inherent flaws of subcritical size, the crack growth rate, V, may be described as [2-4]

$$V = AK^N \tag{1}$$

where K is the stress intensity, and A and N are material/environment constants. If the stress intensity in this equation is rewritten in terms of applied stress, σ, and flaw size, a [2-4] as

$$K = \sigma Y a^{1/2} \tag{2}$$

where Y is a geometric flaw shape factor, the time to failure, t_f, under constant applied stress, σ_A, can be expressed as [2-4]

$$t_f = B\sigma_i^{N-2}\sigma_A^{-N} \tag{3}$$

where

$$B = \frac{2}{AY^2 K_{\text{Ic}}^{N-2}(N-2)} \tag{4}$$

and σ_i and K_{Ic} are the inert strength and critical stress intensity factor, respectively. By expressing the inert strength in terms of its measured failure probability distribution, Eq 3 can be written in the form

$$\ln t_f = \ln B + \left[\frac{N-2}{m_i} \right] [\ln \ln \left(\frac{1}{1-F} \right) + m_i \ln \sigma_{0i}] - N \ln \sigma_A \tag{5}$$

where F is the failure probability and m_i and $\ln \sigma_{0i}$ are the Weibull modulus and scaling parameter, respectively, of the inert strength distribution. Hence, a lifetime prediction or design diagram [2,3,5-7] can be established for the material being considered for any applied stress and failure probability if the inert strength distribution and crack growth parameters, N and B, are known.

Alternatively, the fatigue data can be used to design a proof test which establishes a maximum initial flaw size in the ceramic component if the applied proof load is rapidly removed after it reaches the desired level. Assuming a maximum initial flaw size after proof testing, a relationship between minimum time to failure and proof stress, σ_P, can be expressed as [2]

$$t_f(\text{min}) = B\,\sigma_P^{N-2}\sigma_A^{-N} \tag{6}$$

For structures that are subject to varying levels of applied stress over time, the amount of crack extension occurring during the time at each stress level can be calculated by

$$\left[\frac{a_i}{a_f}\right]^{(2-N)/2} = 1 + \frac{1}{B}\left[\frac{Y}{K_{\mathrm{Ic}}}\right]^{N-2}\left[\frac{\sigma_A^N}{a_f^{(2-N)/2}}\right]t_R \tag{7}$$

where a_i and a_f are the initial and final flaw sizes and t_R is the duration of the applied stress σ_A for each region. Using this relation, it is possible to calculate the necessary proof stress to ensure a maximum initial flaw size so that failure does not occur in less than the minimum desired lifetime.

Experimental

Materials

The specimens evaluated in this investigation were fabricated from a single pour of Corning 9658 machinable glass ceramic. The microstructure of this material consists of a glass matrix with uniformly distributed, randomly oriented mica crystals (platelets) which were homogeneously nucleated in a fluorine-containing parent glass. The bulk mechanical properties and chemical composition of this glass ceramic have been reported [8]. Three groups of specimens were prepared: dynamic fatigue (DF), 125 specimens; static fatigue I (SF-I), 120 specimens; and static fatigue II (SF-II), 80 specimens. Specimens were cut with a diamond-impregnated wafering blade to the nominal dimensions of 6.35 by 0.95 by 0.32 cm. The DF and SF-II specimens were broken in the as-cut condition while the SF-I specimens were ground parallel with a 180-grit resin-bonded diamond spindle prior to testing. The exposure to moisture of all the specimens was normalized through the following process: 24-h, 500°C heat treatment in flowing dry nitrogen; 24-h immersion in distilled water; and storage for a minimum of one week at 50% relative humidity and 24°C prior to testing.

The fracture surfaces and microstructure of the specimens were observed using optical and scanning electron microscopy. A thin film (~ 300 Å) of aluminum evaporated onto the specimen surfaces enabled a qualitative visual characterization of the relative size of the mica platelets in the material. Using this technique, material containing platelets larger than 40 to 50 μm could be identified. Posttest examinations revealed that approximately half of the SF-I specimens failed in regions of large microstructure. These specimens will be referred to as the subgroup SF-IA. Similar microstructural variations were not observed in the other specimen groups.

Inert Strength Measurements

The strength of the glass ceramic specimens measured in the absence of sub-critical crack growth (Region III of the typical $K-V$ diagram [9]) was experimentally determined using instrumented impact testing in a dry nitrogen atmosphere ($<3\%$ relative humidity, 24°C). A pendulum-type tensile impact tester (Custom Scientific Instruments, Inc., Arlington, N.J.), modified for three-point bending (span $= 5.08$ cm), was used to apply an average stress rate of $(2.08 \pm 0.68) \times 10^5$ MPa/s. Strength measurements conducted at even greater stress rates and on wet specimens resulted in values statistically equivalent to measurements made at 2.08×10^5 MPa/s [10]. Therefore, the crack velocities induced at this stress rate are high enough to preclude stress corrosion and hence inert strengths were indeed measured. Strength data were fitted to a two-parameter Weibull distribution [11–13] with the failure probability calculated as

$$F = \frac{n - 0.5}{N} \tag{8}$$

where n is the rank of each failure stress and N is the total number of specimens evaluated.

Dynamic Fatigue Testing

Dynamic fatigue testing [10] was conducted in three-point bending (span $= 5.08$ cm) on a Universal Testing machine (Instron Corporation, Canton, Mass.). Four groups of 25 specimens were tested at stress rates of 0.35, 3.5, 35.0, and 170 MPa/s in ambient air at 60 to 70% relative humidity and 24°C. The median failure strengths, $\hat{\sigma}_f$, were used in the following relation, derived from Eqs 1 and 2 for a constant stress rate, $\dot{\sigma}$

$$\ln \hat{\sigma}_f = \frac{\ln \dot{\sigma}}{N + 1} + \frac{\ln B (N + 1) + (N - 2) \ln \hat{\sigma}_i}{N + 1} \tag{9}$$

to obtain estimates for the crack growth parameters N and B. The median inert strength, $\hat{\sigma}_i$, from impact testing was used to evaluate B.

Static Fatigue Testing

The static fatigue dead-weight-loaded four-point bend fixture shown in Fig. 1 has an inner span of 1.7 cm and an outer span of 5.08 cm. Three groups of 30 SF-I specimens were loaded to 79.3, 82.7, and 86.2 MPa, and the median time to failure was taken as the average time between the 15th and 16th failures. Similarly, the median times to failure were measured for three groups of 20 SF-II specimens loaded to 72.4, 75.8, and 79.3 MPa. All the tests were conducted in air at 23°C and 40 to 60% relative humidity. Using the median times to failure, the crack growth parameters were estimated from the slope and intercept established by Eq 5.

Results and Discussion

The inert strength values of the DF, SF-I, and SF-II specimens obtained from the instrumented impact test were fit to a two-parameter Weibull distribution and are shown in Fig. 2 plotted as a function of failure probability. The apparent deviation from linearity exhibited by these curves may be indicative of two partially concurrent flaw populations [12,13]. The variety of types of failure-initiating defects (edge, surface, or volume flaw) does not account for this behavior, since observations of fracture surfaces indicated that most failures originated on the surface, at or very near an edge. However, the observed

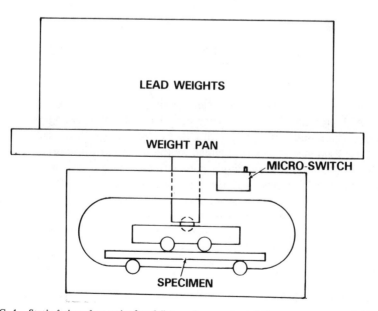

FIG. 1—*Static fatigue four-point bend fixture; inner span = 1.7 cm; outer span = 5.08 cm.*

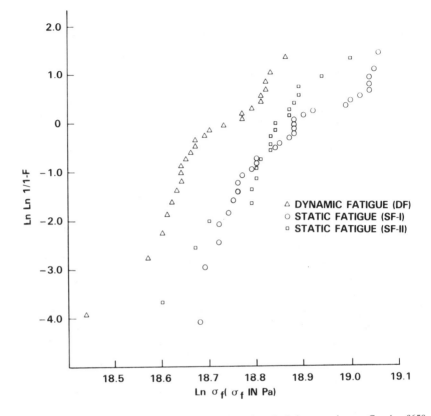

FIG. 2—*Inert strength distributions for dynamic and static fatigue specimens: Corning 9658 machinable glass ceramic. Instrumented impact test: average stress rate = 2.08 × 10⁵ MPa/s.*

variance in the microstructure platelet size may influence the failure distributions. Although all of the specimens contained failure-initiating edge flaws, only some contained strength-limiting flaws associated with the larger microstructure. Evidence of this is provided by the failure distribution for the SF-I specimens (22 out of 30 were larger microstructure specimens), which exhibits the greatest deviation from linearity. Previous work completed on this material revealed a strong correlation between the microstructure size and modulus of rupture. Therefore, the nonlinearity of these distributions may be attributed to the nonuniformity of the specimen microstructure. Since the correlation coefficient (R^2) from linear least-squares analyses of ln σ versus ln ln $1/1 - F$ for each distribution exceeded 0.85, the two-parameter Weibull distribution was used. The values of the Weibull modulus m and scaling parameter ln σ_0 for each distribution, estimated from the regression slope and intercept, are summarized in Table 1, together with the median inert strengths, $\hat{\sigma}_i$.

The details of the dynamic fatigue stress rate testing have been reported

TABLE 1—*Results of inert strength measurements—Corning 9658 machinable glass ceramic three-point bend impact test.*[a]

Specimens	N (number of specimens)	$\hat{\sigma}_i$ (MPa)	m	$\ln \sigma_0$ (σ_0 in Pa)	R^2
DF	25	134.1 ± 2.7	11.9 ± 2.4	18.74	0.93
SF-I	30	158.8 ± 3.5	9.9 ± 1.8	18.92	0.86
SF-IA	22	148.8 ± 2.3	16.8 ± 3.6	18.84	0.92
SF-II	20	153.8 ± 3.6	11.4 ± 2.6	18.88	0.90

[a]Stress rate = 2.08×10^5 MPa/s (±1 standard deviation).

[10], and only the results will be summarized here. The parameters governing crack growth were estimated to be $N = 29.4 \pm 4.0$ and $B = 5.41 \times 10^{10}$ Pa²-s. This value for N is greater than N values reported for many silicate glasses ($N = 15$ to 19) [2,4,5] and may be attributed to an increased fracture surface energy caused by deflections resulting from unfavorably oriented mica platelets in the glass ceramic microstructure [8,14]. These results indicate that this glass ceramic material is only moderately resistant to stress corrosion–induced subcritical crack growth compared with pyroceram glass ceramics with N values reported from 55 to 116 [15]. Hence, delayed failure must be considered if the material is to be used in load-bearing structures. Therefore, design diagrams were constructed from the experimentally determined parameters using Eq 5.

In Fig. 3 is a design diagram, based on the dynamic fatigue data, in which the probability of failure ranges from 0.1 to 90.0%. This information was used to establish a maximum allowable design stress for a ten-year ($\ln t_f$(s) = 19.6) lifetime. Experimental uncertainty in the parameters in Eq 5 (N, B, m_i, σ_0) was used to account for the uncertainty in the maximum design stress using a statistical analysis [16–18]. Weibull edge flaw scaling laws [11,13] (previously established to be applicable [10]) were used to reduce the allowable stress estimated from the design diagram for the smaller test specimens to that acceptable for the actual, larger structures. Finally, a maximum allowable design stress, at 90% confidence, of 16.1 MPa was estimated for a ten-year lifetime at 0.001 probability of failure. This represents approximately 15% of the modulus of rupture value reported for the material by the manufacturer (measured in four-point flexure) and illustrates the effect of delayed-failure mechanisms.

To provide another estimate of the crack growth parameters while more closely modeling the service conditions of the actual structures, static fatigue of this glass ceramic was investigated. The median times to failure under applied load are shown in the $\ln \hat{t}_f$ versus $\ln \sigma_A$ plot in Fig. 4 for the two groups of static fatigue specimens (SF-I and SF-II). Also included in this figure are the results from the Subgroup SF-IA large microstructure specimens. The applied stresses have been converted to equivalent three-point bend stresses [12]. The SF-II specimens were stressed at lower levels than the SF-I specimens because

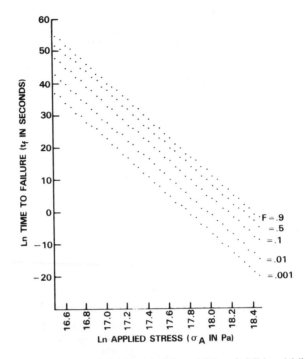

FIG. 3—*Design diagram for 0.001, 0.01, 0.1, 0.5, and 0.9 probabilities of failure based on dynamic fatigue data:* N = 29.4, B = 5.4 × 10^{10} Pa^2-s.

the SF-II specimens exhibited weaker inert strengths. A one-tailed *t*-test for equality of means of the median stress values revealed that the median strengths are significantly different at greater than 95% confidence in all cases. This difference in inert strengths is attributed to the different initial flaw sizes which resulted from the two methods of specimen fabrication. Visual, scanning electron microscopy, and dye penetrant inspection confirmed that grinding resulted in a finer surface finish, especially with respect to edge roughness. The reduced initial flaw sizes of the ground (SF-I) specimens would be reflected by an increase in the time to failure at a given applied stress in relation to the time to failure of the saw-cut (SF-II) specimens. Therefore, the relative shift between the SF-I and SF-II curves in Fig. 4 is not unexpected.

The regression lines of ln $\hat{t}_f$ versus ln σ_A shown in Fig. 4 all have correlation coefficients greater than 0.83. This indicates that the assumed crack growth model (Eq 1) adequately describes the delayed failure mechanisms active in this glass ceramic. The crack growth parameters, N and B, were estimated from the results of the regression analysis and median inert strengths using Eq 5. These values are compiled for each static fatigue specimen group together with the dynamic fatigue results in Table 2. The static fatigue crack growth parameters indicate greater resistance to subcritical crack growth than

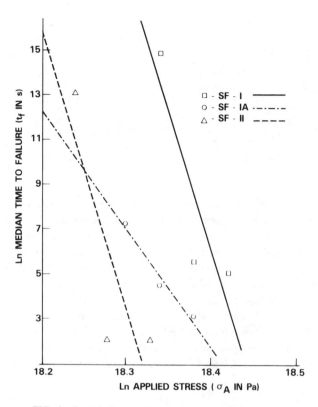

FIG. 4—*Static fatigue median time to failure measurements.*

predicted from dynamic fatigue testing. Although the data obtained are not sufficient to explain the observed differences in N and B values, the longer times to failure may indicate the existence of a threshold stress or fatigue limit for this material. However, this would conflict with evidence indicating that erroneously high N values are obtained in dynamic fatigue for materials exhibiting fatigue limits [19]. Therefore, an investigation similar to one reported for a borosilicate glass [20] is currently underway to determine if a fatigue limit for this glass ceramic exists.

The values for N estimated from the static fatigue data ($N = 50$ for large microstructure specimens, SF-IA; $N = 120$ for SF-I and SF-II) suggest that the resistance to crack growth exhibited by this material is dependent on the size of the microstructure. It is difficult to isolate this effect because the variance in microstructure was observed only after testing. As previously noted, the flexure strength of this glass ceramic increases with decreasing microstructure size, and this trend is attributed to increased numbers of deflections of the propagating crack by the mica platelets as the mica platelet size decreases [8]. Similarly, the resistance to crack growth may be reduced by a decrease in the

TABLE 2—*Crack growth parameters estimated from fatigue testing for Corning 9658 machinable glass ceramic (±1 standard deviation).*

Specimens	Total Number of Specimens	N	$\ln B$ ($B \ln Pa^2$-s)
DF	100	29.4 ± 3.7	24.8 ± 1.7
SF-I	90	117.8 ± 1325	−13.6 ± 675
SF-IA	45	50.0 ± 25.8	18.8 ± 12.4
SF-II	60	121.4 ± 374	−24.6 ± 210

number of crack deflections in the large microstructure material. Since all of the SF-I specimens were machined together, it is assumed that they contain similar flaw populations. Therefore, the SF-1A large microstructure specimens with $N = 50$ have a smaller flaw-size-to-microstructure-size ratio than the SF-I specimens with $N = 118$. Thus, the apparent values estimated for N may be a function of the ratio of the flaw size to microstructure size. Such a dependency was established [21] for several glass ceramics fracture toughness values, which increased as the flaw-size-to-microstructure-size ratio increased.

Design diagrams based on N and B values derived from static fatigue testing of large microstructure specimens, SF-IA, are shown in Fig. 5 for probabilities of failure ranging from 0.1 to 90%. As predicted by the values for the crack

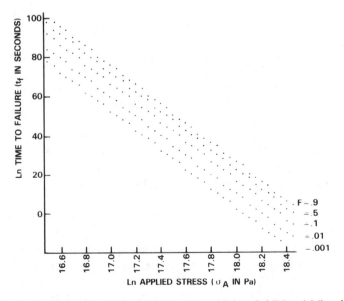

FIG. 5—*Design diagram for 0.001, 0.01, 0.1, 0.5, and 0.9 probabilities of failure based on static fatigue data (SF-IA):* $N = 50.0$, $B = 1.5 \times 10^8$ Pa^2-s.

growth parameters, the times to failure for a given applied stress is seen to increase with respect to the dynamic fatigue design diagram (Fig. 3). A statistical analysis [16–18] of the experimental uncertainty in the estimates of N and B enabled confidence intervals to be assigned to the static fatigue design diagram. The 90% confidence interval is shown in Fig. 6 for the 0.50 failure probability level where $N = 50.0 \pm 25.8$ and $\ln B = 18.80 \pm 12.35$ Pa²-s (SF-IA data). Uncertainty in the times to failure greatly increases as the applied stress considered deviates from the applied stresses used in the static fatigue tests (that is, $\ln \sigma_A = 18.3 - 18.4$). The calculated uncertainty is inversely dependent on the range of applied stresses experimentally investigated, which was narrow for these static fatigue tests. Therefore, the usefulness of the design diagram lifetime predictions was limited by the extent of uncertainty, as illustrated by the confidence interval shown. The 0.50 failure probability level is being discussed for two reasons: this was the level at which the data were taken, and the error associated with extrapolation to lower failure probabilities was unmanageable. However, the crack growth parameters based on the SF-IA data proved to be the most conservative (predicted the greatest amount of crack growth) for establishing a proof stress for the glass ceramic beams.

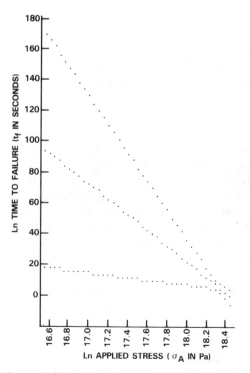

FIG. 6—*An 0.5 failure probability design diagram with 90% confidence interval based on median time to failure static fatigue data (SF-IA):* $N = 50.0$, $B = 1.5 \times 10^8$ Pa²-s.

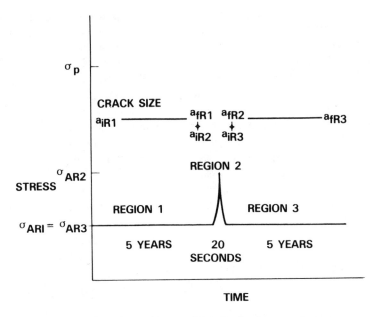

FIG. 7—*Anticipated stress/time conditions for the glass ceramic structures.*

The proof-stress analysis is based on the stress/time conditions anticipated for the ceramic structure. The three stress regions depicted in Fig. 7 represent the five years before launch, launch, and the desired five-year service life. A final flaw size can be estimated from the stress in Region 3, σ_{AR3}, and Eq 2 with K_{Ic} experimentally determined, using double torsion testing, to be 2.27 ± 0.18 MPa-m$^{1/2}$. Substituting this final flaw size for Region 3, a_{fR3}, into Eq 7 with $N = 50.0$, $B = 1.46 \times 10^8$, $t_R = t_{R3}$, and $\sigma_A = \sigma_{AR3}$, the initial flaw size for Region 3, a_{iR3}, can be estimated. Setting the final flaw size in Region 2, a_{fR2}, equal to a_{iR3}, Eq 7 can again be used to calculate the initial flaw size for Region 2, a_{iR2} (with $\sigma_A = \sigma_{AR2}$ and $t_R = t_{R2}$). Similarly, the maximum initial flaw size for Region 1, a_{iR1}, can be estimated (Eq 7), and a corresponding stress, σ_P, can be established from Eq 2. Uncertainty in this flaw size was accounted for by a strength degradation model that enabled application of error propagation laws [22]. Each beam that survives loading to this proof stress in a dry nitrogen atmosphere (with rapid release upon reaching σ_P) will contain flaws no larger than a_{iR1}. This flaw size, under the given loading conditions, will not propagate to failure within the desired lifetime of the glass ceramic structure.

Summary and Conclusions

The susceptibility of a machinable glass ceramic to stress corrosion–induced delayed failure was investigated using dynamic and static fatigue testing. Instrumented impact testing was used to measure inert strengths. Design diagrams for this material, relating lifetime as a function of applied stress and failure probability, were constructed from the results of fatigue and impact testing. The crack growth parameters estimated from static fatigue testing were observed to be dependent on the size of the microstructure of the material. Proof-stress levels were established to preclude failure during the anticipated lifetime stress conditions for the ceramic structures. It was concluded that the most conservative estimates for the crack growth parameters were $N = 50.0 \pm 25.8$ and $\ln B$ (B in Pa^2-s) $= 18.8 \pm 12.4$. The wide variance in these values results from the narrow range of applied stresses investigated.

Acknowledgments

The authors express their gratitude to R. Hunkeler for preparing the specimens used in the study, J. Jellison and D. Kolos for the optical and grain size analyses, R. Marriott, who performed all scanning electron microscopy, and N. Helmold, who designed the static fatigue clock system. Also, the helpful discussions with S. Freiman, J. Mecholsky, W. Viehmann, and C. Staugaitis are gratefully acknowledged.

References

[1] Wiederhorn, S. M., "Subcritical Crack Growth in Ceramics," in *Fracture Mechanics of Ceramics*, Vol. 2, R. C. Bradt, D. P. H. Hasselman, and F. F. Lange, Eds., Plenum, New York, 1974, pp. 613-646.
[2] Ritter, J. E., Jr., "Engineering Design and Fatigue Failure of Brittle Materials," in *Fracture Mechanics of Ceramics*, Vol. 4, R. C. Bradt, D. P. H. Hasselman, and F. F. Lange, Eds., Plenum, New York, 1978, pp. 667-686.
[3] Ritter, J. E. and Meisel, J. A., "Strength and Failure Predictions for Glass and Ceramics," *Journal of the American Ceramic Society*, Vol. 59, Nos. 11-12, 1976, pp. 478-481.
[4] Evans, A. G. and Johnson, H., "The Fracture Stress and Its Dependence on Slow Crack Growth," *Journal of Materials Science*, Vol. 10, 1975, pp. 214-222.
[5] Jakus, K., Coyne, D. C., and Ritter, J. E., "Analysis of Fatigue Data for Lifetime Predictions for Ceramic Materials," *Journal of Materials Science*, Vol. 13, 1978, pp. 2071-2080.
[6] Wiederhorn, S. M., Evans, A. G., Fuller, E. R., and Johnson, H., "Application of Fracture Mechanics to Space-Shuttle Windows," *Journal of the American Ceramic Society*, Vol. 57, No. 7, 1974, pp. 319-323.
[7] Humenik, J. H. and Ritter, J. E., "Susceptibility of Alumina Substrates to Stress Corrosion During Wet Processing," *Journal of the American Ceramic Society*, Vol. 59, No. 12, 1980, pp. 1205-1211.
[8] Chyung, K., Beall, G. H., and Grossman, D. G., "Microstructures and Mechanical Properties of Machinable Glass-Ceramics," in *Electron Microscopy and Structure of Materials*, G. Thomas, Ed., University of California Press, Berkeley, Calif., 1972, pp. 1167-1194.
[9] Chandon, H. C., Bradt, R. C., and Rindone, G. E., "Dynamic Fatigue of Float Glass," *Journal of the American Ceramic Society*, Vol. 61, Nos. 5-6, 1978, pp. 207-210.
[10] Smyth, K. K. and Magida, M. B., "Dynamic Fatigue of a Machinable Glass-Ceramic," *Journal of the American Ceramic Society*, Vol. 66, No. 7, 1983, pp. 500-505.
[11] Robinson, E. Y., "Estimating Weibull Parameters for Materials," NASA TM 33-580, JPL, National Aeronautics and Space Administration, Jet Propulsion Laboratory, Pasadena, Calif., 1972.
[12] Johnson, C. A., "Fracture Statistics in Design and Application," Report No. 79 CRD 212, G. E. Corporate Research and Development, Schenectady, N.Y. 1979.

[13] Bansal, G. K. and Duckworth, W. H., "Effects of Specimen Size on Ceramic Strengths," in *Fracture Mechanics of Ceramics*, Vol. 3, R. C. Bradt, D. P. H. Hasselman, and F. F. Lange, Eds., Plenum Press, New York, 1978, pp. 189–204.

[14] Pletka, B. and Wiederhorn, S., "Subcritical Crack Growth in Glass Ceramics," in *Fracture Mechanics of Ceramics*, Vol. 4, R. C. Bradt, D. P. H. Hasselman, and F. F. Lange, Eds., Plenum, New York, 1978, pp. 745–754.

[15] Cook, R. R., Lawn, B. R., and Anstis, G. R., "Fatigue Analysis of Brittle Materials Using Indentation Flaws, Part 2: Case Study on a Glass-Ceramic," *Journal of Materials Science*, Vol. 17, 1982, pp. 1108–1116.

[16] Ritter, J. E., Bandyopadhyay, N., and Jakus, K., "Statistical Reproducibility of the Dynamic and Static Fatigue Experiments," *Bulletin of the American Ceramic Society*, Vol. 60, No. 8, 1981, pp. 799–806.

[17] Jacobs, D. F. and Ritter, J. E., "Uncertainty in Minimum Lifetime Predictions," *Journal of the American Ceramic Society*, Vol. 59, Nos. 11–12, 1976, pp. 481–487.

[18] Wiederhorn, S. M., Fuller, E. R., and Mandel, J., "An Error Analysis of Failure Prediction Techniques Derived from Fracture Mechanics," *Journal of the American Ceramic Society*, Vol. 59, Nos. 9–10, 1976, pp. 403–411.

[19] Minford, E. J. and Tressler, R. E., "Effects of a Threshold Stress Intensity for Crack Growth and Flaw Blunting on Dynamic Fatigue Curves," paper presented at the 85th Annual Meeting of the American Ceramic Society, Chicago, Ill., 1983.

[20] Hayashi, K., Easler, T. E., and Bradt, R. C., "A Fracture Statistics Estimate of the Fatigue Limit of a Borosilicate Glass," paper presented at the 85th Annual Meeting of the American Ceramic Society, Chicago, Ill., 1983.

[21] Mecholsky, J. J., "Fracture Mechanics Analysis of Glass-Ceramics," *Advances in Ceramics*, Vol. 4, American Ceramic Society, 1982, pp. 261–276.

[22] Ritter, J. E., "Assessment of Reliability of Ceramic Materials," in *Fracture Mechanics of Ceramics*, Vol. 5, R. C. Bradt, A. G. Evans, D. P. H. Hasselman, and F. F. Lange, Eds., Plenum, New York, 1983, p. 227.

Sheldon M. Wiederhorn,[1] *Stephen W. Freiman,*[1]
Edwin R. Fuller, Jr.,[1] *and Herbert Richter*[2]

Effect of Multiregion Crack Growth on Proof Testing

REFERENCE: Wiederhorn, S. M., Freiman, S. W., Fuller, E. R., Jr., and Richter, H.,
"Effect of Multiregion Crack Growth on Proof Testing," *Methods for Assessing the Structural Reliability of Brittle Materials, ASTM STP 844*, S. W. Freiman and C. M. Hudson,
Eds., American Society for Testing and Materials, Philadelphia, 1984, pp. 95-116.

ABSTRACT: The effect of subcritical crack growth on proof testing is examined. Crack
velocity curves obtained by fracture mechanics techniques are used to predict theoretical
strength distributions for the specimens that survive proof testing. These theoretical distributions are then compared with experimental distributions obtained on soda lime silica
glass slides. The comparison reveals a strong sensitivity of the proof test results to the exact
position and shape of the crack growth curve. Minor changes in the crack growth curve
result in major shifts in position and shape of the strength distribution curves after proof
testing. The importance of crack geometry and specimen configuration to crack growth
behavior, and hence to the strength distribution, is emphasized.

KEY WORDS: crack growth, cracks, fracture mechanics, glass, proof testing, strength,
Weibull statistics, structural reliability, brittle materials

In order to improve the reliability of ceramic materials in structural applications, proof testing is often used as a method of controlling the strength distribution. Weak components are broken by the proof stress, leaving a population of components with a minimum value of the strength, which in the
absence of subcritical crack growth, is equal to the stress used during the proof
test. Thus, proof testing truncates the strength distribution and guarantees
that all components will be at least as strong as the minimum value. Because
of its simplicity, proof testing has been used successfully as a basis of design in
many structural applications [1-10].

Potential problems in the application of proof testing occur when subcritical

[1]Group leader, ceramic engineer, and physicist, respectively, Inorganic Materials Division,
National Bureau of Standards, Washington, D.C. 20234.
[2]Senior research scientist Fraunhofer-Institut fur Werkstoffmechanik, D-800 Frieburg, West
Germany.

crack growth accompanies proof testing. An example of this is seen in a recent study of soda lime silica glass tested in air or water [11]. It was shown that, under certain test conditions, the proof test did not truncate the strength distribution. Furthermore, the test results did not fully agree with the theory that was used to explain them [12]. The study suggested the need for additional research on the effect of multiregion crack growth on proof testing. It is because of this need that the present study was conducted. The study is similar to one recently completed by Ritter et al [13], and the results complement those obtained by those authors.

Experimental Procedure

Heptane was chosen as a test medium for this study because of its low affinity for water. At normal temperatures and pressures heptane contains only ~ 50 ppm water, which corresponds to a partial pressure of water equal to 30% relative humidity. The net effect of the low concentration of water is to suppress Region II crack growth behavior to a velocity of $\sim 10^{-6}$ m/s but to leave the curve for Region I crack growth virtually unchanged. The range of the crack growth curve in Region II, 0.5 MPa·m$^{1/2}$ to 0.7 MPa·m$^{1/2}$, is greater than that obtained in most other environments, and therefore is expected to enhance the effect of multiregion crack growth on proof testing.

The proof test specimens were unannealed microscope slides with dimensions of 75 by 25 by 1 mm. The test conditions and the number of specimens used in each test are given in Table 1. All the specimens were loaded to the proof load in four-point bending at a rate of 33 MPa/s. As the equipment was automated to reduce the load as soon as the proof test load was reached, the hold time at the proof test load was less than 0.1 s and is assumed to be zero in the present study. Based on crack growth data reported in Ref 14, and preliminary strength measurements in heptane, the unloading rates were selected to yield the maximum range of predicted effects of proof testing on the strength distribution curve after proof testing. Following the proof test, the specimens were loaded to failure at a loading rate of 33 MPa/s without changing their position in the test fixture. Thus, the stress distribution within the test specimen was the same when strength was measured as it was during the proof test.

Crack velocity data were obtained using two fracture mechanics configurations. Applied-moment, double-cantilever beam specimens (75 by 25 by 1 mm) were used to collect crack velocity data over the velocity range 10^{-10} to 10^{-3} m/s; crack velocities were measured by direct observation of the crack tip as a function of time [15]. Single-edge, notched tensile specimens (130 by 25 by 2 mm) [16] were used to collect data over the velocity range 10^{-9} to 100 m/s. In the velocity range 10^{-9} to 10^{-7} m/s, velocities were measured by monitoring the crack tip visually. Higher velocity measurements were made by the use of an acoustic pulse technique described in detail elsewhere [17]. A sonic pulse was used to mark the fracture surface so that the position of the crack front was es-

TABLE 1—*Test conditions for soda lime silica glass proof tested in heptane.*[a]

Loading Rate, MPa/s	Unloading Rate, MPa/s	Proof Load, MPa	Number Tested	Proof Test Survivors
33	strength tests	...	628	...
33	165	103	673	417
33	3.3	100.5	630	179
33	0.33	90.5	497	143

[a]All strength measurements were made by four-point bending using an inner span of 19 mm and an outer span of 64 mm.

tablished as a function of time. Velocities were then determined after completion of the experiment by microscopic examination of the crack surface.

To compare the crack velocity data with the strength data, a computer code was designed to calculate the breaking stress of the specimens after the proof test. The computer code is a modified version of the one discussed in Ref *12*. Input data for the code are (1) Weibull parameters, S_0 and m, obtained in the test environment (heptane); (2) a v-K_I plot expressed as a piecewise power-law fit to the v-K_I data; (3) the fracture mechanics parameters, K_{Ic}, K_{Iscc}, and Y (Y is defined by the equation $K_I = Y\sigma\sqrt{a}$ where a is the characteristic crack size, σ the applied stress, and K_I the stress intensity factor);[3] and (4) the proof test parameters (stressing rate during loading and unloading, proof stress, and hold time, which is assumed to be zero in the present experiment).

The computer code first calculates the initial distribution of strength from the initial distribution obtained in heptane and the input data enumerated previously.[4] This procedure provides a common basis for comparing crack growth data and proof test data and was used by Ritter et al [*13*] in their study. After the initial distribution has been determined, the computer code determines the effect of the proof test cycle on this distribution by first calculating the degree of strength degradation (that is, the amount of crack growth) during the proof test cycle, and then calculating the breaking stress in heptane of the specimens that survived the proof test cycle. The results of this analysis can then be compared directly with experimental strength distribution data obtained in heptane. The main output of the computer code is a graphical representation in which Weibull probability coordinates are used to display the strength distribution after proof testing. On command, the computer code was capable of plotting the initial strength distribution in heptane, the calculated initial distribution, and the distribution after proof testing.

[3]The value of K_{Ic} used in the present study was 0.75 MPa·m$^{1/2}$. The reason for selecting this value is discussed in the Appendix.

[4]The initial values of m and S_0 determined in this study are listed in Table 3 of the Appendix. This calculation is needed because subcritical crack growth occurs during the strength tests. The initial distribution, calculated from the computer code, is an idealized distribution that would be determined if strength measurements could be made without the occurrence of subcritical crack growth.

The value of Y (1.5) used in the present paper is representative of the maximum stress intensity factor at a semielliptical surface crack with a minor to major axis ratio of ~ 0.6 [18]. This ratio was observed in the present study for indentation cracks in glass surfaces (see Fig. 7). We assume that Y remains constant as the surface crack increases in size. This assumption permits us to use a one-dimensional model to determine the strength distribution after proof testing. For real cracks, the shape of the crack, and consequently the value of Y, probably changes during crack growth. Although this aspect of crack growth is important, the simpler treatment adopted here should permit us to identify the principal factors that influence strength degradation during proof testing.

Results

Crack Growth Data

The results of the crack velocity measurements are given in Fig. 1a. Approximately 100 data points were collected by each of the two techniques. Data for the single-edge, notched (SEN) specimens were spread evenly over the velocity range 10^{-7} to 10^2 m/s; the double-cantilever beam (DCB) data were spread evenly over the velocity range 10^{-10} to 10^{-3} m/s.[5] Thus, the data from the SEN specimens are concentrated in a higher range of crack velocities than the data from the DCB specimens. Only five data points were obtained on SEN specimens at velocities of $< 10^{-7}$ m/s.

Because the two sets of velocity data overlap over much of the data range, differences between the data are shown more clearly by fitting each set of data piecewise with straight lines (Fig. 1b).[6] When this is done, the crack growth data obtained by the SEN technique appear to lie at slightly higher values of K_I than those obtained by the double-cantilever beam technique. A small lateral translation of the two sets of data by approximately 0.05 MPa·m$^{1/2}$ brings them into close coincidence and, for practical purposes, makes the two sets of data indistinguishable over most of the data range.

The difference between the two sets of data in Region II possibly results from the difference in thickness of the specimens used in the two techniques. The greater thickness of the SEN specimens increases the diffusion path for water from the bulk to the crack tip environment and suppresses the Region II data to lower crack velocities. This type of data has been reported earlier by Richter [17].

At velocities that exceed $\sim 10^{-4}$ m/s, the slope of the v-K_I curve is observed to decrease, almost as if there were a second Region II behavior. At $K_I \sim 0.93$

---

[5]The abbreviations SEN and DCB are changed to SE(T) and DB(Mx) in the ASTM Standard Terminology Relating to Fracture Testing (E 616-82). A discussion of the terminology can be found in the *1983 Annual Book of ASTM Standards*.

[6]A summary of the piecewise linear fit to the crack growth data is given in Table 4 of the Appendix.

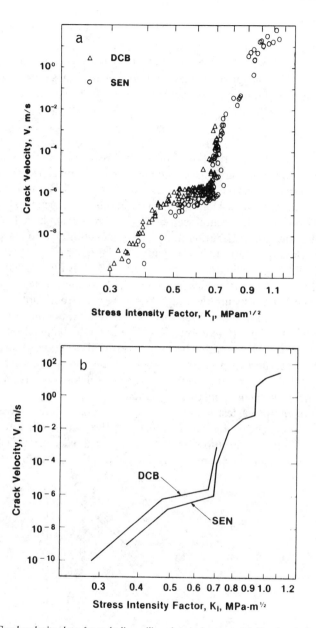

FIG. 1—*Crack velocity data for soda lime silica glass in heptane.* (a) *Crack velocity data collected by the applied-moment, double-cantilever techniques and the single-edge, notch technique.* (b) *Piecewise, straight-line fit of the crack growth data shown in* (a). *The curve marked SEN represents the single-edge, notch data; the curve marked DCB represents the double-cantilever data.*

$MPa \cdot m^{1/2}$ the crack velocity is observed to increase precipitously by about one to two orders of magnitude. This sudden increase in velocity has been attributed to cavitation of the fluid near the crack tip [19], whereas the decrease in slope of the v-K_1 curve at velocities that exceed $\sim 10^{-4}$ m/s is probably due to viscous drag of the fluid on the fracture surfaces [20].

Strength Data

The initial strength distribution obtained by breaking 628 microscopic slides in heptane is shown in Fig. 2. Expressed as a two-parameter Weibull distribution, the parameters for the regression line are $S_0 = 112$ MPa for the intercept, and $m = 7$ for the slope. Despite the large number of specimens tested, the curve still exhibits experimental scatter about the regression line. The applicability of the strength distribution to the specimens subjected to the proof test can be checked qualitatively by comparing the number of specimens broken during the loading portion of the proof test cycle with those predicted from Fig. 2. This procedure is permissible since the loading rate was the same for all four-point bend tests in this study. As can be seen from Table 2, the measured and predicted number of failures differ by 1% for the medium and slow unloading rates. The difference of 7% (34 versus 41% failures) for the rapid unloading rate can be accounted for by a 4% error in either the proof test level, or S_0 for the proof-tested specimens. In view of this small difference in strength, we feel justified in using the Weibull parameters determined from Fig. 2 to describe the underlying strength distributions.

Results of the proof test study (Fig. 3) indicate the occurrence of a large number of failures below the proof stress for all three unloading rates. For the medium unloading rate, approximately 42% of the specimens failed below the proof stress when they were reloaded to failure. For the rapid and slow un-

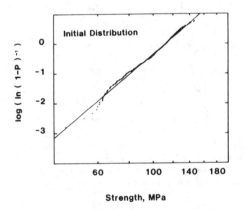

FIG. 2—*Initial strength distribution of 628 microscope slides broken in four-point bending, in heptane:* $S_0 = 112$; m = 7.

TABLE 2—*Test of Weibull distribution ($S_0 = 112$ MPa, m = 7).*

Unloading Rate, MPa/s	Proof Load, MPa	Fraction Broken During Loading	
		Measured	Predicted
165	103.0	0.34	0.41
3.3	100.5	0.38	0.37
0.33	90.5	0.21	0.20

loading rates, 16% of the survivors failed below the proof stress upon reloading to failure. This decrease in strength is consistent with theoretical expectations, since subcritical crack growth is expected to weaken all the specimens that survive the proof test. The specimens that are weakened most by the proof test are those with strengths that lie closest to the proof test stress prior to testing. The weakest specimen after testing had strengths of 74, 61, and 50% of the proof stress for the rapid, medium, and slow unloading rates, respectively. This finding agrees with that reported by Ritter et al [13], who noted that the strength degradation of specimens just passing the proof test is greater as the unloading rate is decreased.

The strength distribution curves shown in Fig. 3 give little indication that the proof test truncated the distribution curves. Only the fastest unloading rate gave evidence of truncation. However, for this curve, a large number of specimens broke at strengths that were less than the proof stress. Strength distribution curves for the medium and slow unloading rates differ only slightly from the initial strength distribution. Because of the large number of specimens tested, however, we believe that the small changes in shape and position of strength distribution curves shown in Fig. 3 are statistically significant. As will be shown, these changes in shape and position can be explained by the occurrence of subcritical crack growth during the proof test.

Comparison of Crack Velocity Data with Proof Test Data

Strength Distribution Curves

A comparison of the strength distribution curves with the curves calculated from the v-K_1 data is given in Fig. 4. The straight lines in this figure represent the initial distribution of specimen strengths obtained in heptane. The data points were selected from Fig. 3 to indicate the location of the data; not all of the data points are given. The curves in Fig. 4 are the theoretical estimates of the strength distribution after proof testing, determined from the crack velocity data in Fig. 1b.

For the rapid unloading rate, the two theoretical curves predict sharply truncated strength distributions. Considering this prediction, few strengths should

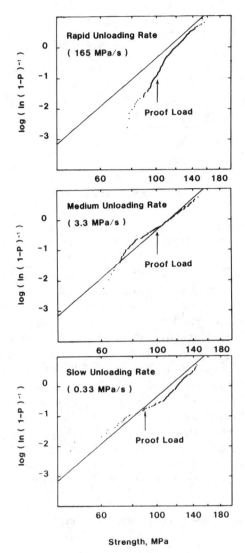

FIG. 3—*Strength distribution after proof testing in heptane:* (a) *rapid (165 MPa/s) unloading rate;* (b) *medium (3.3 MPa/s) unloading rate;* (c) *slow (0.33 MPa/s) unloading rate.*

have been measured with values of less than ~ 100 MPa. However, in contrast to this expectation, the experimental curve was not sharply truncated; approximately 16% of the specimens broke at stresses less than 100 MPa. Although the experimental data fall along the theoretical curves at strength levels greater than 100 MPa, there is a low-strength "tail" in the data that is not explained by either theoretical curve. This "tail" was also observed by Ritter et al [13] in their recent study of soda lime silica glass proof tested in heptane.

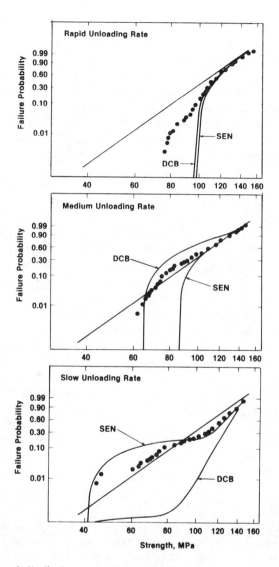

FIG. 4—*Strength distribution curves after proof testing in heptane, calculated from* v-K_I *data shown in Fig. 1:* (a) *rapid (165 MPa/s) unloading rate;* (b) *medium (3.3 MPa/s) unloading rate;* (c) *slow (0.33 MPa/s) unloading rate.*

The strength distribution for the rapid unloading rate (Fig. 4) differs considerably from the distribution obtained in an earlier study in which the proof testing was conducted in dry nitrogen gas [*11*]. In the earlier study, the experimental data closely approximated the theoretical curve. The difference in results between studies is probably not attributable to water in the environment. In both studies the unloading rate was sufficiently fast that the strength degradation during the proof test should have been controlled by the "water-free" portions of the crack growth curves. It is possible that the difference in results is due to physical rather than chemical aspects of the two environments. Viscosity, for example, is much greater for liquids than for gases; therefore viscous drag by the liquid on the crack surfaces could retard the motion of the crack and permit some specimens to survive that would otherwise have failed during the proof test. Clearly, additional studies are needed to confirm this possibility and, more generally, to explore differences in the effect of liquids and gases on the strength of materials. From a practical point of view, however, the results from the present study in heptane and the earlier one in nitrogen gas suggest that proof testing in liquids may not be a wise procedure, because the strength distribution curve may not be sharply truncated when the test is performed in a liquid.

In the case of the medium and slow unloading rates (Fig. 4) the theoretical curves depart significantly from the experimental data; the departure is much greater than would be expected from the scatter of the experimental data. Furthermore, the two theoretical curves in these figures show a surprisingly large divergence considering the small difference between the crack velocity curves in Fig. 1. These differences between the experimental data and the theoretical curve, on the one hand, and the two theoretical curves, on the other, suggest that the strength distribution curves after proof testing are extremely sensitive to the exact location of the curves that describe crack motion. As can be seen by comparing Fig. 1 with the medium and slow unloading rate curves of Fig. 4, small changes in the position of the v-K_I curve result in very large differences in the predicted position of the strength distribution curve after proof testing. This extreme sensitivity of the strength distribution to the crack velocity also explains the differences between the experimental and theoretical strength distribution curves in Fig. 4.

Maximum Crack Velocity During Proof Testing

The sensitivity of the strength distribution curves to the exact shape and position of the crack velocity curves can be understood in terms of the maximum velocity achieved by the critical flaw in the course of the proof test. During a proof test, strength loss occurs at all levels of applied stress (above the fatigue limit) as a consequence of subcritical crack growth. The strength loss is most severe where the crack velocity is highest. Thus, the high-velocity regimen of the v-K_I diagram is most important for determining the distribution of strengths after proof testing. In fact, because most crack growth occurs at the maximum crack velocity, strength degradation should be related to the maximum velocity achieved in the course of the proof test: the higher the velocity the greater the amount of strength degradation.

To examine the relationship between the maximum crack velocity during proof testing and the strength distribution curve, the maximum crack velocity was calculated for each of the points used to plot the theoretical curves shown in Fig. 4. The maximum velocity was determined and then printed next to each point on the theoretical curve to indicate the maximum velocity during the proof test. An example of the type of data generated is shown in Fig. 5 for the slow unloading rate. (Results for the fast and medium unloading rates are similar to Fig. 5 and are not discussed here.) As can be seen from Fig. 5, the range of maximum velocities was relatively narrow for both sets of curves, covering approximately two orders of magnitude (10^{-6} to 10^{-8} m/s) for the SEN specimens and one order of magnitude (10^{-7} to 10^{-8} m/s) for the DCB specimens. Regardless of the unloading rate, or the v-K_I data, the narrow velocity range was common to each of the strength distribution curves obtained in the present study. Thus, only a small portion of the v-K_I plot is effective in establishing the position and shape of the strength degradation curve after proof testing. This point is illustrated in Fig. 6, which shows the range of crack velocities that was effective in establishing the strength degradation curves shown in Fig. 5. Moving the v-K_I plot by a small amount dramatically changes the portion of the v-K_I curve that determines the strength distribution after proof testing. For example, in Fig. 5 the shape of the DCB curve is determined primarily by Region I crack growth behavior. Lowering the Region II portion of the v-K_I curve slightly, as has been done for the SEN specimen, permits a large fraction of the specimens to achieve velocities that are characteristic of Region II crack growth behavior. Because the slope of the strength distribution curve depends on the slope of the crack growth curve [12], this change in crack growth behavior dramatically alters the shape of the strength distribution curve after proof testing.

The general shape and appearance of the strength distribution curves in

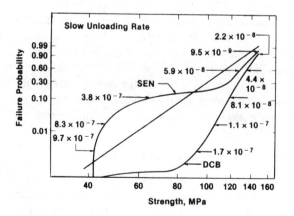

FIG. 5—*Maximum velocities (in metres per second) achieved during proof testing printed next to calculated strength distribution curves. The maximum crack velocities and the calculated strength distribution curves were obtained from the v-K_I data shown in Fig. 1: slow (0.33 MPa/s) unloading rate.*

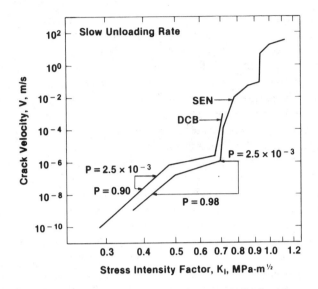

FIG. 6—*Range of crack velocities that determine the strength distribution curves for the slow (0.33 MPa/s) unloading rate.*

Fig. 5 are a direct consequence of the fact that these curves were calculated from the crack velocity data in Fig. 1. Both the crack growth and strength curves are multiregioned. Because of the importance of the maximum velocity as a determinant for strength, each segment of the strength distribution curve is related to a segment of the crack growth curve. However, the two curves are inverted in relation to one another—the high-velocity region of the crack growth curve maps onto the low-probability region of the strength distribution curve, and vice versa. Furthermore, the slopes of corresponding segments of the two curves are also related: the slope, m, of the strength distribution curve is given by $n - 2$, where n is the slope of the crack velocity curve [12]. This one-to-one relationship between the two curves plays an important role in determining the sensitivity of the strength distribution curve to the crack velocity curve, especially when only a narrow range of crack velocity data is relevant to the strength distribution curve, as is the case in the present study.

Correlation Between Theory and Experiment

In view of the foregoing discussion, the lack of correlation between the experimental and theoretical strength distributions can be understood in terms of the high sensitivity of the strength distribution curve to the exact shape and position of the crack growth curve. Any physical or chemical process that disturbs the position of this v-K_1 curve, or prevents an accurate determination of the curve, can affect the position and shape of the strength distribution curve and thus result in a lack of agreement between the theoretical and experimental strength distribution curves. Some of the causes of poor agreement between the theoretical and experimental distribution curves will now be discussed. As will be seen, the causes are of two types: experimental errors that prevent an accurate determination of the "true" v-K_I curve, and physical or chemical effects that alter the behavior of small surface cracks in such a way that their response to applied stress is significantly different from that obtained on large macroscopic cracks. Of the two causes, the latter are more fundamental to our understanding of crack growth and strength degradation. However, in practice both can be equally important for the determination of accurate strength distribution curves.

Experimental Uncertainty

The experimental uncertainty associated with the crack growth data is illustrated in Fig. 1 by the small difference in position of the two v-K_I curves and the random scatter of the data of each curve. The difference between the two curves (Fig. 1a) represents a systematic uncertainty in measurement that has its origin in the technique used to make the measurement. This type of error can be reduced only by more thorough theoretical and experimental investigations of the two crack velocity techniques to trace the origin of the measurement difference. The random error in Fig. 1 is given by the scatter of data about their mean value for any given value of K_I or v. As can be seen from this figure, the scatter in K_I ranges from ~3 to 10% of K_I, whereas the error in v ranges from ~0.2 to ~1 order of magnitude in crack velocity, depending on the values of v and K_I and the technique used to determine the v-K_I plot. These errors are typical of those that have been reported previously in the literature for crack velocity studies on glass.

With regard to the present experiment, the experimental scatter of the data in Fig. 1 is sufficient to preclude a determination of the true v-K_I curve with sufficient accuracy to predict the position and shape of the strength distribution curve after testing. As a consequence of this finding, it is suggested that more accurate measurement techniques are needed to determine crack growth behavior. Without more accurate methods of measurement, one should be wary of using crack growth data to predict proof testing results. However, even with better measurement techniques there may be more basic reasons that strength distribution curves cannot be determined accurately from crack growth data. These reasons relate to the equivalence of crack growth data from large and small cracks.

Large versus Small Cracks

Although it is usual to assume that the growth of microscopic size cracks in ceramic materials is described by crack growth data obtained by macroscopic, fracture mechanics techniques, there are good physical reasons why this assumption may not be valid. In other words, small cracks that control the strength of glass may not behave in the same way as large cracks used in fracture mechanics studies. Three factors that may contribute to the difference in crack growth behavior are (1) residual stresses due to plastic deformation at the origin of the surface cracks, (2) blunting of small cracks due to stress corrosion, and (3) differences in the v-K_I curve resulting from geometric differences between large and small cracks.

Residual Stresses

The effect of residual stresses on the strength of glass was first discussed by Marshall and Lawn [21,22]. The effect results from the fact that surface cracks in glass are usually formed by mechanical damage during the grinding and polishing of glass or by mechanical contact with hard substances after the glass has been produced. Residual stress fields always form at points of mechanical contact as a consequence of microplastic deformation at contact asperities. Unless the residual stresses are relieved by annealing, they contribute to the stress field that forces cracks to grow during a proof test. Therefore, these stresses should be taken into account to determine the strength distribution after proof testing. As the crack grows, the effect of residual stresses gradually becomes less important; the effect is least for the weakest specimens that survive the proof test. Hence, the effect of residual stresses on the strength distribution is expected to be most pronounced for the high-probability portion of the stress distribution. In this paper it is assumed that residual stresses played a small role in the strength degradation process. This assumption is supported by the fact that the proof test results in Fig. 3 are very similar to those obtained by Ritter et al on annealed specimens [13]. However, to fully understand the role of residual stresses on proof testing, additional studies are clearly needed.

Crack Blunting

By applying transition rate theory to ceramic materials, Charles and Hillig [23] suggested the existence of a static fatigue limit caused by crack-tip blunting in materials such as glass. At stresses above the fatigue limit, surface flaws sharpen and grow, whereas at stresses below the fatigue limit, crack-tip blunting and crack arrest occur. The Charles–Hillig theory is consistent with all known experimental data on the strength of glass. Recently, Michalske, using fracture mechanics techniques, has demonstrated a fatigue limit in soda lime silica glass at a value of $K_I \sim 0.25$ MPa $\cdot$ m$^{1/2}$ [24]. The fatigue limit was identified by a hysteresis in the crack velocity curve and by microstructural features that were left on the face of a crack that was repropagated after having been blunted by stress corrosion. Resharpening of cracks in soda lime silica glass occurs almost instantaneously when the stress intensity factor exceeds

0.425 MPa $\cdot$ m$^{1/2}$. If the stress intensity for resharpening falls within the range of K_I that is effective in determining the strength distribution after proof testing, then crack blunting can have a significant effect on the shape of the strength distribution curve after proof testing. Our studies on this point are quite limited; however, they do show that crack blunting is more important for those proof tests that were conducted using slow unloading rates than fast ones. In addition, our studies show that the strongest specimens are not affected at all by the proof test, whereas the weakest ones suffer the same degree of strength degradation as would occur if there were no static fatigue limit. Additional studies to clarify the role of crack blunting on proof testing should include the establishment of resharpening kinetics and the incorporation of these kinetics into the strength treatment.

Crack Geometry

Effects of environment on both strength and crack growth rate have been studied by a number of investigators. Studies on soda lime silica glass indicate relatively good agreement between strength measurements and crack growth measurements, when measurements are conducted either in water, where Region I crack growth is dominant, or in an inert environment such as dry nitrogen, where Region III crack growth is dominant. For both techniques the stress corrosion susceptibility, n, was found to be similar regardless of the technique used to make the measurement.[7] Therefore, it is unlikely that the behavior of large cracks differs significantly from that of small ones when only Region I or Region III crack growth is involved in degrading the strength [25]. The same conclusion, however, may not be valid when Region II crack growth occurs.

Crack growth in Region II has been studied by a number of investigators. Wiederhorn [26] and Schönert et al [27] were the first to attribute Region II crack growth to an environment-limited transport of water to the crack tip. Schönert et al demonstrated that the crack front changed its shape as the crack moved from Region I to Region II, and again when it moved from Region II to Region III. When studies are conducted with through cracks in flat specimens (edge-notched or double-cantilever beam specimens), the portion of the crack front in the center of the specimen always leads the portion of the crack front that intersects the specimen surface in Region I or III. In Region II, however, the reverse is true: the portion of the crack front near the surface always leads that in the center.

Quackenbush and Frechette [28] and Richter [29] confirmed the findings of Schönert et al and attributed crack growth in Region II to a combined effect of Region I and Region III crack growth behavior. Since water is freely available at the specimen surface, crack motion is enhanced by water, whereas

---

[7]When residual stresses affect the crack growth behavior, a correction must be made for the effect of these stresses on the stress corrosion susceptibility, n. However, even in this case, the value of n for crack growth can be related to that obtained from fatigue data [22,23].

away from the specimen surface the crack tip is starved for water because of the longer diffusion path, and crack motion is retarded. In specimens with through cracks, the cracks move as a unit in Region II, the center portion of the crack being pulled along by portions near the surface. As the applied stress intensity factor in Region II is increased, the crack velocity is observed to increase, but not as rapidly as in Regions I or III. Furthermore, as K_I is increased, more of the crack perimeter behaves as if it were depleted of water, and finally when the outer portion of the crack is completely depleted of water, it, too, enters Region III and the crack perimeter regains its normal shape. Thus, Region II represents a transition in crack growth behavior between Regions I and III; depending on the availability of water at the crack tip, different portions of the crack front behave somewhat independently as the crack moves.[8]

The effects just described also occur for surface cracks that are responsible for the failure of strength specimens. The effect is shown in Fig. 7, which illustrates successive positions of a surface crack that was propagated to failure in normal heptane. The crack was formed by the use of a diamond pyramidal indenter; the velocity was measured by the use of a sonic pulse generator. The crack position is indicated by ripple marks on the fracture surface, each set of marks indicating successive positions of the crack as it grows larger. In Fig. 7a the ripple marks closest to the indentation (halfway between the two arrows) represent Region I crack growth, whereas those furthest from the indentation represent Region III crack growth. The arrows mark the transition between Region II and Region III crack growth. As can be seen from Fig. 7b, the shape of the crack in Region II differs from the normal semielliptical shape that is characteristic of Region I or Region III crack motion. As in the case of the through cracks discussed previously, the change in shape can be attributed primarily to water depletion at portions of the crack that are distant from the surface. Because of water depletion, these portions of the crack move slower than those near the surface where the water is plentiful. Thus, the shape of the surface crack is modified by the water content of the environment near the crack perimeter. Because of the difference in geometry between small surface cracks and the large cracks that are typical of fracture mechanics specimens, physical constraints on the crack front are different for the two types of cracks.[9] Therefore, crack growth in Region II will probably not be the same for the two types of cracks even though crack growth in Region II represents a transition in behavior between Regions I and III for both crack types. The differences between the theoretical and experimental strength distribution curves in Fig. 4 may be the result of these geometric considerations. Substantiation of this belief, however, will require the solution of three-dimensional elasticity/diffu-

---

[8]In an earlier paper, Varner and Frechette [30] suggest that crack arrest occurs periodically in the central portion of the crack in Region II (for studies conducted in nitrogen gas). A later study by Richter [29] (conducted in air) and by Quackenbush and Frechette [28] (conducted in decane) gave no indication of crack arrest in Region II. The results of the present study on indentation cracking also give no indication of crack arrest in Region II.

[9]For example, K_I for an elliptical crack can vary by as much as 50% along the crack perimeter.

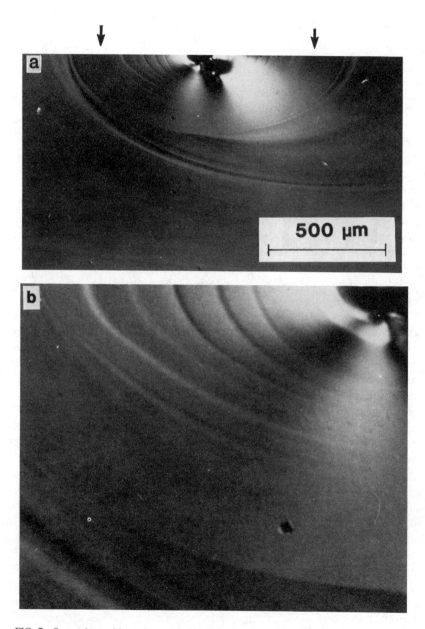

FIG. 7—*Successive positions of a surface crack that was propagated to failure in heptane:* (a) *the arrows mark the transition between Region II and Region III crack growth;* (b) *higher magnification of* (a) *showing the shape of the crack fronts in Region II.*

sion boundary value problems to determine exactly how crack growth occurs for surface cracks. This task is beyond the scope of the present paper.

Summary

This paper presents a critical analysis of the applicability of crack growth data to proof testing as a means of improving the reliability of structural ceramic materials. In particular, the paper deals with the question of predicting strength distribution curves from crack growth data obtained by standard fracture mechanics techniques. To deal with this question, both strength data and crack growth data were obtained on soda lime silica glass specimens that were tested in liquid heptane. Crack growth data were obtained by the applied-moment, double-cantilever beam (DCB) technique and by the single-edge, notch tension (SEN) technique. Although there were systematic differences between the two sets of crack velocity data, a partial overlap of the data was observed over most of the data range. Considering that two different test techniques were used to collect the data, the two sets of crack growth data were considered to be consistent with the degree of scatter that has been reported for crack growth data.

Proof test experiments were conducted on soda lime silica glass microscope slides that were subjected to four-point bending. Strength distribution curves after proof testing were obtained using three different unloading rates during the proof test cycle. Truncation of the strength data was obtained only for specimens subjected to a rapid (165 MPa/s) rate of unloading. Specimens that were subjected to either slow (0.33 MPa/s) or medium (3.3 MPa/s) rates of unloading gave curves that hardly differed from the initial strength distribution curve. Furthermore, the three strength distribution curves obtained from the proof tested population of glass slides could not be predicted from the crack growth data obtained in the present experiment.

In the course of our discussion and analysis, we concluded that the position and shape of the strength distribution curve after proof testing was highly sensitive to the exact position and shape of the crack growth data. We were able to show that slight shifts in the position or shape of the v-K_1 plot resulted in substantial changes in the position and shape of the predicted strength distribution curve. As a consequence, any physical or chemical process that disturbs or changes the position of the crack growth curve or prevents an accurate determination of the crack growth curve will cause the strength distribution curve determined from crack growth data to differ considerably from the one determined experimentally. Specific causes of the differences are (1) the experimental uncertainty inherent in obtaining v-K_1 data and (2) physical and chemical effects that alter the behavior of small surface cracks so that their response to applied stress differs significantly from the response of large macroscopic cracks typical of fracture mechanics specimens.

From a practical point of view our strength data suggest that proof testing in an "inert" liquid may not be a wise procedure because only a modest degree of truncation is observed (only at high unloading rates) and because after proof testing many specimens have strengths that lie below the proof test level. Based on these observations and results from an earlier study, which included proof test studies in dry nitrogen gas, we suggest that proof testing be conducted in dry, gaseous environments and that rapid rates of unloading be used to ensure structural reliability. When this procedure is followed, satisfactory truncation of the strength distribution can be achieved.

Acknowledgment

This work was supported by the U.S. Office of Naval Research, Metallurgy and Ceramics Program. The technical assistance of Mark Gollup and Meers Oppenheim and the helpful technical discussions with B. R. Lawn are gratefully acknowledged.

APPENDIX

The Selection of K_{Ic}

In developing the computer code used in the present study, we found it necessary to establish a criterion for failure. A standard, albeit somewhat idealized, failure criterion is to assume that failure occurs when the applied stress, σ, is equal to the strength, S [31]. In fracture mechanics terms, this criterion is equivalent to assuming $K_I = K_{Ic}$ and $d(K_I/da) > 0$ for the most critical crack.

In more practical terms, however, failure by crack growth is determined by a crack velocity criterion: failure occurs when the crack is moving so rapidly that the specimen cannot be unloaded fast enough to avoid failure. The critical velocity for failure depends on the rate of unloading: as the unloading rate is increased, the critical velocity also is increased. This failure criterion is exactly applicable to proof testing, since the number of specimens that pass the proof test is determined by the rate of unloading.

When the v-K_I curve is steep, as it is for soda lime silica glass in nitrogen gas, the two criteria are almost identical. The value of K_{Ic} is insensitive to the crack velocity, and the value selected (within Region III) for K_{Ic} does not affect the truncation that results from proof testing. However, when the v-K_I curve is complicated by physical effects, such as viscous drag or cavitation, as it is for soda lime silica glass tested in heptane, then one must be concerned with the equivalence of the two failure criteria. In particular, it must be determined whether K_{Ic} evaluated by conventional methods can be used to establish a criterion for crack stability for a proof test.

In the present study we used a value of $K_{Ic} = 0.75$ MPa$\sqrt{m}$ as obtained on DCB specimens that were loaded to failure in dry nitrogen gas [32]. To test the validity of this conventional value of K_{Ic}, the v-K_I curve for the SEN specimen (Fig. 1b) was used to determine a strength distribution curve for three values of K_{Ic}: 0.75, 0.90, and 1.10 MPa $\cdot$ m$^{1/2}$. For each of the proof test cycles used in the present study, the strength distribution after proof testing and the number of specimens that broke during the proof test were found to be independent of the value of K_{Ic} used in the program. The reason for this apparent insensitivity of the strength distribution curve to K_{Ic} is the fact that the distribution curve is determined by a relatively narrow range of velocities on the v-K_I plot. As long as K_{Ic} lies at a value that is above this range of velocities, the strength curve will not

TABLE 3—Input data: Weibull parameters.

Condition	S_0	m
Experimental	112	7
Inert: calculated from v-K_1 data		
SEN	123.1	6.55
DCB	130.1	6.28

TABLE 4—Crack growth data.[a]

Condition	Region	Slope, n	$\ell n\ A$
SEN[b]	1	16.59	−233.13
	2	5.07	−81.90
	3	245.66	−3315.52
	4	56.00	−762.88
	5	13.02	−180.91
	6	6.97	−98.23
	7	382.50	−5255.06
	8	16.03	−218.66
	9	6.60	−88.35
DCB	1	17.10	−237.95
	2	4.09	−67.85
	3	125.83	−1700.19

[a] Slopes and intercepts for v-K_1 curves; values are based on pascals as the units for stress.
K_{1c} = 0.75 Mpa·m$^{1/2}$
K_{1scc} = 0.2 MPa·m$^{1/2}$
Y = 1.5
[b] The number of significant figures given in these parameters is included for purposes of calculation only and does not indicate accuracy of the various parameters.

be sensitive to its value. The main effect of changing K_{1c} is to shift the position of the initial strength distribution.[10] Increasing the value of K_{1c} increases the value of S_0 and decreases the value of m for the initial distribution (Table 3). Thus, K_{1c} plays the role of a scaling factor in a proof test. In view of this conclusion, the fracture mechanics value of K_{1c} (that is, 0.75 MPa · m$^{1/2}$) was used for the present study (Table 4).

References

[1] Wiederhorn, S. M., "Reliability, Life Prediction, and Proof Testing of Ceramics," in Ceramics for High Performance Applications, J. J. Burke, A. G. Gorum, and R. N. Katz, Eds., Brook Hill, Chestnut Hill, Mass., 1974, pp. 635–655.
[2] Ritter, J. E., Jr., "Engineering Design and Fatigue Failure of Brittle Materials, in Fracture Mechanics of Ceramics, Vol. 4, R. C. Bradt, D. P. H. Hasselman, and F. F. Lange, Eds., Plenum, New York, 1978, pp. 667–686.
[3] Humenik, J. N. and Ritter, J. E., Jr., "Susceptibility of Alumina Substrates to Stress Corro-

[10] The initial distribution referred to here is the distribution calculated from the input data discussed in the section on experimental procedure.

sion Cracking During Wet Processing," *Bulletin of the American Ceramic Society*, Vol. 59, 1981, p. 1205.

[4] Wiederhorn, S. M., Evans, A. G., and Roberts, D. E., "A Fracture Mechanics Study of the Skylab Windows," in *Fracture Mechanics of Ceramics*, Vol. 2, R. C. Bradt, D. P. H. Hasselman, and F. F. Lange, Eds., Plenum, New York, 1974, pp. 829-841.

[5] Wiederhorn, S. M., Evans, A. G., Fuller, E. R., and Johnson, H., "Application of Fracture Mechanics to Space-Shuttle Windows," *Journal of the American Ceramic Society*, Vol. 57, 1974, pp. 319-323.

[6] Ritter, J. E., Jr., and Wulf, S. A., "Evaluation of Proof Testing to Assure Against Delayed Failure," *Bulletin of the American Ceramic Society*, Vol. 57, 1978, pp. 186-190.

[7] Ritter, J. E., Jr., Sullivan, J. M., Jr., and Jakus, K., "Application of Fracture Mechanics Theory to Fatigue Failure of Optical Glass Fibers," *Journal of Applied Physics*, Vol. 49, 1978, pp. 4779-4782.

[8] Greene, D. G., Ritter, J. E., Jr., and Lange, F. F., "Evaluation of Proof Testing as a Means of Assuring Mission Success for the Space Shuttle Thermal Protection System," *Material and Process Application—Land, Sea, Air, Space*, Vol. 26, Science of Advanced Materials and Process Engineering Series, Society for the Advancement of Material and Process Engineering, Azusa, Calif., 1981, pp. 257-269.

[9] Greene, D. G., Ritter, J. E., Jr., and Lange, F. F., "Fracture Behavior of Low Density Fibrous Ceramic," *Journal of the American Ceramic Society*, Vol. 65, 1982, pp. 141-146.

[10] Ritter, J. E., Jr., "Assessment of the Reliability of Ceramics," in *Fracture Mechanics of Ceramics*, Vol. 5, R. C. Bradt, A. G. Evans, D. P. H. Hasselman, and F. F. Lange, Eds., Plenum, New York, 1983, pp. 227-251.

[11] Ritter, J. E., Jr., Oates, P. B., Fuller, E. R., Jr., and Wiederhorn, S. M., "Proof Testing of Ceramics: I. Experiment," *Journal of Materials Science*, Vol. 15, 1980, pp. 2275-2281.

[12] Fuller, E. R., Jr., Wiederhorn, S. M., Ritter, J. E., Jr., and Oates, P. B., "Proof Testing of Ceramics: II. Theory," *Journal of Materials Science*, Vol. 15, 1980, pp. 2282-2295.

[13] Ritter, J. E., Jr., Jakus, K., Young, G. M., and Service, T. H., "Effect of Proof-Testing Soda-Lime Glass in Heptane," *Journal of the American Ceramic Society*, Vol. 65, 1982, pp. C134-C135.

[14] Freiman, S. W., "Effects of Straight-Chain Alkanes on Crack Propagation in Glass," *Journal of the American Ceramic Society*, Vol. 58, 1975, pp. 339-340.

[15] Freiman, S. W., Mulville, D. R., and Mast, P. W., "Crack Propagation in Brittle Materials," *Journal of Materials Science*, Vol. 8, 1973, pp. 1527-1534.

[16] Richter, H., "Experimentelle Untersuchungen zur Rissausbreitung in Spiegelglass im Geschwindigkeitbereich 10^{-3} bis 5×10^3 mm/s," Institut fur Festkorpermechanik der Fraunhofer-Gesselschaft e.V. Freiburg i. Br. Rosastrasse 9, December 1974.

[17] Richter, H., "Der Einfluss der Probendicke auf die Rissausbreitung in Glas," *Proceedings, Eleventh International Congress on Glass*, Vol. II, Prague, 1977, pp. 447-457.

[18] Irwin, G. R. and Paris, P. C., "Fundamental Aspects of Crack Growth and Fracture," in *Fracture, An Advanced Treatise*, Vol. 3, H. Liebowitz, Ed., Academic Press, New York, 1971, pp. 1-46.

[19] Michalske, T. A. and Frechette, V. D., "Dynamic Effects of Liquids on Crack Growth Leading to Catastrophic Failure in Glass," *Journal of the American Ceramic Society*, Vol. 63, 1980, pp. 603-609.

[20] Wiederhorn, S. M., Freiman, S. W., Fuller, E. R., Jr., and Simmons, C. J., "Effects of Water and Other Dielectrics on Crack Growth," *Journal of Materials Science*, Vol. 17, 1982, pp. 3460-3478.

[21] Marshall, D. B. and Lawn, B. R., "Residual Stress Effects in Sharp Contact Cracking, I. Indentation Fracture Mechanics," *Journal of Materials Science*, Vol. 14, 1979, pp. 2001-2012.

[22] Marshall, D. B., Lawn, B. R., and Chantikul, P., "Residual Stress Effects in Sharp Contact Cracking, II. Strength Degradation," *Journal of Materials Science*, Vol. 14, 1979, pp. 2225-2235.

[23] Charles, R. J. and Hillig, W. B., in *Symposium on Mechanical Strength of Glass and Ways of Improving It*, Florence, Italy, 25-29 September, 1961, pp. 511-527. Union Scientifique Continentale du Verre, Charleroi, Belgium, 1962.

[24] Michalske, T. A., "The Stress Corrosion Limit: Its Measurement and Implications," *Fracture Mechanics of Ceramics*, Vol. 5, R. C. Bradt, A. G. Evans, D. P. H. Hasselman, and F. F. Lange, Eds., Plenum, New York, 1983, pp. 277-289.

[25] Dabbs, T. P., Lawn, B. R., and Kelly, P. L., "A Dynamic Fatigue Study of Soda-Lime Silicate and Borosilicate Glasses Using Small Scale Indentation Flaws," *Physics and Chemistry of Glasses*, Vol. 23, 1982, pp. 58–66.

[26] Wiederhorn, S. M., "Influence of Water Vapor on Crack Propagation in Soda-Lime Glass," *Journal of the American Ceramic Society*, Vol. 50, 1967, pp. 407–414.

[27] Schönert, K., Umhauer, H., and Klemm, W., "The Influence of Temperature and Environment on the Slow Crack Propagation in Glass," in *Fracture 1969*, Proceedings of the Second International Conference on Fracture, Brighton, England, 1969, pp. 474–482.

[28] Quackenbush, C. L. and Frechette, V. D., "Crack-Front Curvature and Glass Slow Fracture," *Journal of the American Ceramic Society*, Vol. 61, 1978, pp. 402–406.

[29] Richter, H., "Rissfrontkrümmung und Bruchflächenmarkierung im Übergangsbereich der Bruchgeschwindigkeit," *Glastechnische Berichte*, Vol. 47, 1974, pp. 146–147.

[30] Varner, J. R. and Frechette, V. D., "Fracture Marks Associated with Transition Region Behavior of Slow Cracks in Glass," *Journal of Applied Physics*, Vol. 42, 1971, pp. 1983–1984.

[31] Kapur, K. C. and Lamberson, L. R., "Reliability in Engineering Design," Wiley, New York, 1977.

[32] Wiederhorn, S. M., "Fracture Surface Energy of Glass," *Journal of the American Ceramic Society*, Vol. 52, 1969, pp. 99–105.

DISCUSSION

G. Quinn[1] *(written discussion)*—You have demonstrated that lifetime predictions are extremely sensitive to the exact position of the *K-V* curve. Slight shifts in crack velocity data can translate into huge changes in lifetime predictions. Do you think analysis methods based on crack velocity measurements will ever become widely applicable for life prediction purposes?

S. M. Wiederhorn, S. W. Freiman, E. R. Fuller, Jr., and H. Richter (authors' closure)—For structural ceramic components, lifetime prediction at room temperature is based on the concept of crack growth from preexisting flaws. However, since data for the prediction of lifetimes of components can be obtained either from strength or from crack growth measurements, it is unclear at the present time which type of measurement will eventually be used as the basis for lifetime prediction. Problems that arise when crack growth data are used for lifetime prediction occur both in the accuracy with which data can be obtained and in the modeling of the motion of small flaws that control strength, crack motion being affected by such factors as crack-tip blunting, residual stresses, crack geometry, and microstructure. Until all of these effects are fully understood and correctly modeled, caution must be exercised when using crack growth data to predict the lifetime of ceramic components. We are gradually learning to deal with these problems, and perhaps when they are fully understood, crack growth data will be used as a reliable means of predicting lifetime.

[1]U.S. Army–AMMRC, Watertown, Mass. 02172.

Gerald G. Trantina[1]

Fracture Mechanics Analysis of Defect Sizes

REFERENCE: Trantina, G. G., **"Fracture Mechanics Analysis of Defect Sizes,"** *Methods for Assessing the Structural Reliability of Brittle Materials, ASTM STP 844*, S. W. Freiman and C. M. Hudson, Eds., American Society for Testing and Materials, Philadelphia, 1984, pp. 117–130.

ABSTRACT: Stress rupture and crack velocity data must be used to predict the lifetime of structures. However, to use stress rupture data, the large amount of scatter characteristic of brittle materials must be treated, and to use crack velocity data the initial defect size must be known. Since the time-dependent failure of many brittle materials involves propagation of a crack from an initial defect to a critical size at which stress rupture occurs, crack velocity and stress rupture data can be combined to predict the initial defect size. This technique has been developed and used to provide (1) statistical interpretations of stress rupture data as a size distribution of critical defects and (2) probability of failure predictions of structures, including the size effect. An expression for the stress rupture curve which includes the initial defect size is produced by integrating an expression for the crack velocity curve that includes the threshold stress intensity factor. The scatter is treated along lines of constant initial stress intensity factor, so that the scatter is independent of rupture life. This approach is applied to literature data for soda lime glass in water. An initial defect size is inferred for 261 data points, and the Weibull distribution is used to display the data in terms of the inverse square root of the initial defect size. The defect size distribution for soda lime glass is predicted, and an example of the failure prediction of a structure is included where the size effect is considered.

KEY WORDS: fracture mechanics, defect sizes, stress rupture, crack growth rate, scatter, size effect, Weibull distribution, stress intensity factor, soda lime glass, structural reliability, brittle materials

Techniques must be developed to assess the structural reliability of brittle materials. A number of investigators have explored various aspects of statistical design with brittle materials [1–8]. Also, the stress rupture behavior (time-to-failure for constant stress loading) of brittle materials, particularly glass, has been widely studied [9–13]. The stress rupture failure mechanism

[1]Manager—CAD technology, General Electric Company, Plastics Business Group, Pittsfield, Mass. 01201.

of many brittle materials involves the propagation of a crack from an initial defect to a critical size where failure is defined. An analysis of this process involves three types of material data—crack growth rates, stress rupture, and initial defect size. Any two of these three types of data can be used to predict the third.

In the analysis presented in this paper, a relationship is fitted to crack growth rate data. For an initial defect size, the relationship is integrated to produce a stress rupture curve. An initial defect size is thus inferred for a stress rupture data point that falls on the curve. Therefore, a distribution of critical initial defect sizes can be predicted from a set of stress rupture data. It is presumed that the variation or scatter in stress rupture data is dependent only on the variation of the initial defect size and not on material variability or other sources. A similar approach has been applied to cyclic crack growth and fatigue data [14]. The objectives of this paper are to describe the approach and demonstrate its application to soda lime glass by determining both the size distribution of critical initial defects and the size effect of the average defect size. By relating stress rupture life to critical defect sizes in test specimens, critical defect size distributions in components can be predicted, and allowable material defect sizes can be based on required component life.

Crack Growth Analysis

The crack velocity, da/dt, is controlled by the stress intensity factor, K, which is a measure of the magnitude of the stress field at the crack tip and is computed from the applied stress, σ, and the crack length, a,

$$K = Y\sigma\sqrt{a} \tag{1}$$

where Y is a geometric factor that accounts for the geometry of crack and structure. The crack growth rate can be expressed as a power law

$$\frac{da}{dt} = B\left(\frac{K}{K_{th}} - 1\right)^n \tag{2}$$

where B, n, and K_{th} are constants. The threshold stress intensity factor, K_{th}, is added as the third constant to account for the curvature of a typical crack growth rate curve. For many materials, da/dt becomes very small at small values of K, indicating an apparent threshold stress intensity factor

$$K_{th} = Y\sigma\sqrt{a_{th}} \tag{3}$$

where a_{th} is the threshold crack length for an applied stress, σ. For $a > a_{th}$ cracks will propagate, and for $a < a_{th}$ cracks will not propagate.

The crack growth rate expression (Eq 2) can be used to predict the time to failure, t_f, assuming the following:

1. The failure process consists entirely of the propagation of a crack from an initial size to a critical size.

2. The shape of the crack remains constant during propagation of the crack, and its size is always small in relation to the distance to any free surface.

3. The crack driving force is characterized by the elastic stress intensity factor.

4. The applied stress is constant during crack propagation.

The crack growth rate expression can be integrated by substituting Eqs 1 and 3 into Eq 2 and separating the variables

$$\int_{a_i}^{a_f} \frac{da}{(\sqrt{a} - \sqrt{a_{th}})^n} = \frac{BY^n \sigma^n}{K_{th}^n} \int_0^{t_f} dt \tag{4}$$

The initial crack length, a_i, is defined in terms of the initial stress intensity factor

$$K_i = Y\sigma\sqrt{a_i} \tag{5}$$

and the final crack length, a_f, is defined by the critical value of K

$$K_c = Y\sigma\sqrt{a_f} \tag{6}$$

The assumed shape of the crack is a semicircular surface flaw of length a where $Y = 1.12(2/\pi)(\sqrt{\pi})$. Integrating Eq 4 yields the time to failure

$$t_f = \frac{2K_{th}^n}{BY^n \sigma^n} \left[\frac{(\sqrt{a_f} - \sqrt{a_{th}})^{2-n}}{2-n} + \sqrt{a_{th}} \frac{(\sqrt{a_f} - \sqrt{a_{th}})^{1-n}}{1-n} \right.$$

$$\left. - \frac{(\sqrt{a_i} - \sqrt{a_{th}})^{2-n}}{2-n} - \sqrt{a_{th}} \frac{(\sqrt{a_i} - \sqrt{a_{th}})^{1-n}}{1-n} \right] \tag{7}$$

The stress rupture curve is thus a function of the three constants from the crack growth rate expression, B, n, K_{th}, the fracture toughness, K_c, the crack geometry, Y, and the initial crack length, a_i. Of these parameters, the initial crack length is probably the most difficult to determine and has the most sensitive effect on the stress rupture life. Another form of Eq 7 that is useful in understanding the parameters that control stress rupture life is

$$t_f = \frac{2K_{th}^n}{BY^2\sigma^2}\left[\frac{(K_c - K_{th})^{2-n}}{(2-n)} + K_{th}\frac{(K_c - K_{th})^{1-n}}{(1-n)}\right.$$
$$\left. - \frac{(K_i - K_{th})^{2-n}}{(2-n)} - K_{th}\frac{(K_i - K_{th})^{1-n}}{(1-n)}\right] \qquad (8)$$

In this expression it can be seen how the initial stress intensity factor, K_i, affects the rupture life. The time to failure is a function of several constants but only two variables, σ and K_i. In the following section the process of inferring a size distribution of critical defects by using Eq 7 and the implications of Eq 8 on scatter in stress rupture data will be discussed.

Analysis of Stress Rupture Data

The scatter in stress rupture data is assumed to arise only from a variation in initial defect size and not from material variability or other sources. A stress rupture curve can be produced by using Eq 7 or 8 and by assuming a set of material properties, B, n, K_{th}, K_c, a crack geometry, Y, and an initial crack size, a_i. Two such curves could be produced for two values of a_i to represent scatter in stress rupture data from two specimens of the same size but with different initial defect sizes, or they could represent average stress rupture curves for specimens with two different sizes where the larger specimens would have, on the average, a larger initial defect size and a shorter time to failure. Treating the scatter or size effect (specimen to specimen or specimen to structure) as variability in rupture life at constant stress results in increasing variability with increasing rupture life. At long lifetimes, the scatter becomes enormous and life prediction is complicated by tests that have been terminated without failure. A more satisfactory method treats scatter and size effects as variability in rupture strength at constant life. This results in only a small increase in the scatter or specimen size effect of the rupture strength with increasing life. Nevertheless, treatment of scatter either at constant life or at constant stress results in variability or scatter which varies with lifetime. The following describes a treatment of rupture data in which the variability is constant throughout the stress rupture curve.

Equation 8 can be expressed as

$$t_f\sigma^2 = f_1(K_i) = f_2(\sigma\sqrt{a_i}) \qquad (9)$$

so that lines of constant K_i have a slope of -0.5 on plots of log σ versus log t_f. A constant value of K_i can result from a small crack and a large stress or a large crack and a small stress (Eq 9). In either case, the initial response to the crack driving force, K, is the same since the starting point K_i, on the crack growth rate curve is the same. The lifetimes are different because $t_f\sigma^2$ is constant (Eq 9). The rupture curves for different a_i values are identical in shape but offset along lines which have a slope of -0.5. The distribution of critical flaw sizes in a material can be pictured as a family of such curves. The relative positions of the curves are identical along all lines with a slope of -0.5; therefore, the scatter or size effect is independent of stress and life when analyzed along such lines. Thus, scatter and size effects should be treated along lines of constant initial stress intensity factor, K_i.

To infer a distribution of critical initial defect sizes from stress rupture data and crack growth rate data, each stress rupture data point, σ, t_f, must be associated with a stress rupture curve. Data for soda lime glass tested in water were used as an example. Crack growth rate data [11] were used to establish the constants in the crack growth rate expression (Eq 2), where $B = 3.2 \times 10^{-6}$ m/s $(1.3 \times 10^{-4}$ in./s), $n = 6$, and $K_{th} = 0.22$ MPa$\sqrt{m}$ $(0.20$ ksi$\sqrt{in.})$. The crack growth rate data for soda lime glass in water are shown in Fig. 1, along with the expression for the crack growth rate curve. The trend of the data indicates that a threshold may exist. The stress rupture data [8] for soda lime glass in water are shown in Fig. 2. The tests were performed at five different stress levels by loading abraded glass slides in four-point bending. A total of 261 tests were performed, but 55 tests were terminated without specimen failure (runouts). The median t_f is indicated in Fig. 2. The curve in Fig. 2 will be described in the next section. Now each stress rupture data point can be associated with a stress rupture curve by iteratively computing a_i until the stress rupture curve passes through the data point. The a_i value is then an estimate of the initial size of the defect which led to the failure of that specimen.

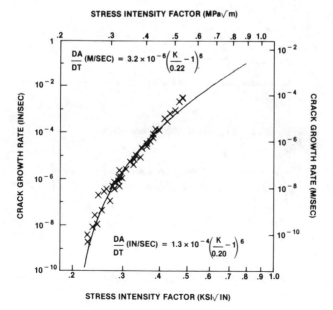

FIG. 1—*Crack growth rate data* [11] *for soda lime glass in water.*

Distribution of Defect Sizes

After the initial sizes of the strength-controlling defects have been computed for a data set, a mathematical representation must be used to statistically treat the size distribution. The two-parameter Weibull distribution function [15] has been chosen

$$P = 1 - \exp\left(-\int \left(\frac{\sigma}{\sigma_0}\right)^m dA\right) \qquad (10)$$

where P is the cumulative probability of failure, m is the Weibull modulus, and σ_0 is a normalizing constant. The integration of $(\sigma/\sigma_0)^m$ is carried out over all elements of surface, dA, if the defects leading to stress rupture failure are located only on the surface. Otherwise, the integration is carried out over the volume. The Weibull modulus is a measure of the scatter or variability in defect sizes where a small value of m indicates a large degree of variability. Accordingly, the Weibull modulus describes the variability or scatter in rupture strength and the sensitivity of the rupture strength to specimen size. Equation 10 can be expressed as

$$P = 1 - \exp\left(-kA\left(\frac{\sigma_{max}}{\sigma_0}\right)^m\right) \qquad (11)$$

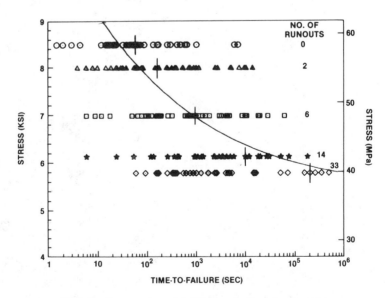

FIG. 2—*Stress rupture data* [8] *for soda lime glass in water.*

where σ_{max} is the maximum σ on the surface and k is a dimensionless constant defined as

$$k = \int^{A} \left(\frac{\sigma}{\sigma_{max}} \right)^{m} \frac{dA}{A} \tag{12}$$

Tensile specimens with uniform stress throughout result in $k = 1$. All other stress states result in $0 < k < 1$. The product of k times A in Eq 11 is often described as the "effective area" and is the area of the specimen which is effectively being stressed at σ_{max}.

Since scatter and size effects should be treated at constant K_i (and therefore constant $\sigma \sqrt{a_i}$), σ in Eq 11 can be replaced by $1/\sqrt{a_i}$. A graphical method of displaying the distribution as a straight-line relationship involves a rearrangement of Eq 11 to

$$\ell n \, \ell n \, \frac{1}{1 - P} = m \, \ell n \, \frac{1}{\sqrt{a_i}} + \text{constant} \tag{13}$$

If one plots $\ell n \, \ell n \, 1/(1 - P)$ versus $\ell n \, 1/\sqrt{a_i}$, the data points should fall along a straight line with a slope equal to the Weibull modulus, m. The probability of failure, P, for each specimen can be estimated as $(n - 0.5)/N$, where n is the ordering number (weakest to strongest or largest to smallest crack length) and N is the total number of specimens.

An example of size distributions of critical surface defects in soda lime glass is shown in Fig. 3. The cumulative probability of failure is displayed versus $1/\sqrt{a_i}$. These five curves correspond to the five levels of stress used in the stress rupture tests described in the previous section. The curves are plotted separately because there are runouts at each stress level except the highest. Since the runouts at each stress level represent the strongest specimens (smallest defect size), they occupy the highest rank but cannot be plotted because they did not fail, and thus an a_i cannot be estimated. This is illustrated in Fig. 3 as an increasing lack of data at the upper part of the curves as the stress level decreases and there are more runouts. The Weibull distribution provides a reasonable fit of the size distribution of critical defects.

A summary of the defect size analysis of soda lime glass is given in Table 1. The important feature to note is the comparison of the defect size distributions for the five stress levels (Fig. 3). These defect distributions should be independent of the stress level. Two measures of the defect distributions are the slope (Weibull modulus) and the median defect size. The results in Table 1 indicate that the Weibull modulus, $m \simeq 9$, and the median defect size, $a_i \cong$ 23 μm (0.0009 in.), are nearly constant. The curve in Fig. 2 is drawn by using this defect size. Further consistency in the average defect size is obtained by considering the inert strength and K_c, both measured in liquid nitrogen. The inert strength is 132 MPa (19.1 ksi) [8], and K_{Ic} is 0.82 MPa$\sqrt{m}$ (0.75

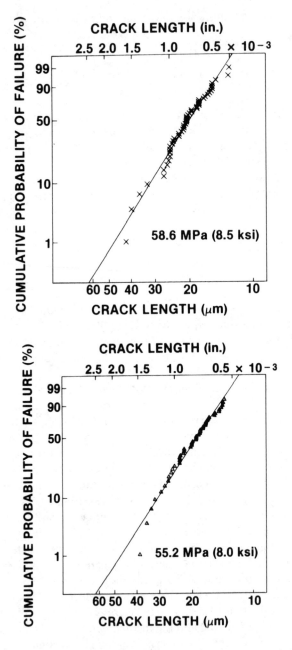

FIG. 3—*Defect distributions as Weibull plots of probability [ℓn ℓn 1/*

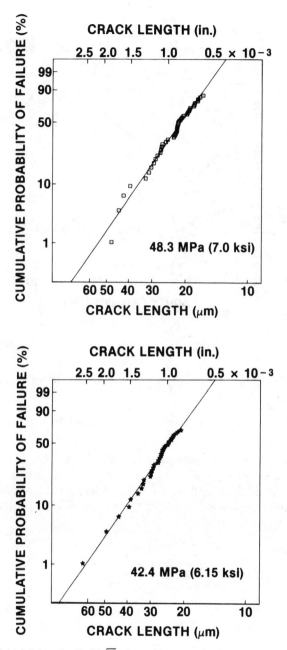

(1 − P)] versus initial defect size ($\ell n \ 1/\sqrt{a_i}$) for various stress levels.

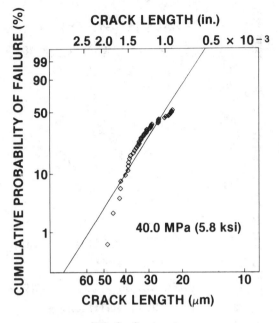

FIG. 3—*Continued.*

ksi$\sqrt{\text{in}}$.) [16]. Using Eq 1 and these values, a defect size of 25 μm (0.001 in.) was calculated.

Size Effect

The size effect in brittle material occurs because the size of the critical defect, on the average, increases with increasing area or volume or with more uniform stress distributions (larger k). This size effect also can occur between specimens and components. From Eq 11 one can show that at constant probability of failure

$$\frac{\sigma_1}{\sigma_2} = \left(\frac{k_2 A_2}{k_1 A_1}\right)^{1/m} \tag{14}$$

Thus, for K_i = constant

$$\frac{\sqrt{a_{i2}}}{\sqrt{a_{i1}}} = \left(\frac{k_2 A_2}{k_1 A_1}\right)^{1/m} \tag{15}$$

The specimen size effect is now given in terms of the square root of the average initial defect size.

TABLE 1—*Summary of stress rupture test results for soda lime glass in water.*

Applied Stress, MPa (ksi)	No. of Specimens Tested	No. of Runouts	Median Time to Failure, s	Median Defect Size, μm (in.)	Weibull Modulus
58.6 (8.5)	47	0	60	21 (0.81 × 10⁻³)	9.7
55.2 (8.0)	47	2	261	19 (0.76)	9.3
48.3 (7.0)	47	6	1 002	23 (0.89)	9.1
42.4 (6.15)	47	14	10 131	25 (0.98)	8.8
40.0 (5.8)	73	33	260 832	24 (0.94)	9.8

A pressurized, soda lime glass cylinder is used as an example of the failure prediction of a structure where the size effect is included. Equation 15 is used to translate the defect size distribution for the glass slides tested in four-point bending to pressurized glass cylinders. The outer span for the four-point bend tests was 5.08 cm (2.0 in.), and the inner span and the specimen width was 2.54 cm (1.0 in.). Since the specimens were abraded, only surface flaws, and thus surface area, are considered. Also, only the tensile surface is used because flaws do not tend to propagate in compressive stress fields. The area between the inner loading points (6.45 cm² − 1.0 in.²) is uniformly stressed

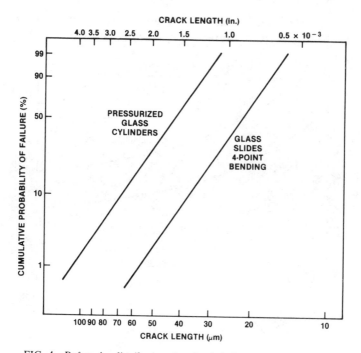

FIG. 4—*Defect size distributions for abraded glass slides and cylinders.*

($k_1 = 1$). The area between the inner and outer loading points (6.45 cm² − 1.0 in.²) has a stress that varies linearly from 0 to the maximum stress. For a linearly varying stress distribution Eq 12 is used to show that $k_1 = 1/(m + 1)$. Since $m = 9$ for abraded soda lime glass, $k_1 = 0.1$. These two effective areas, $k_1 A_1$, are summed and used in Eq 15 along with $k_2 A_2$ for the pressurized cylinder, where $k_2 = 1$ since the stress is nearly uniform in a thin-walled cylinder. The cylinder is 7.62 cm (3.0 in.) long and 2.54 cm (1.0 in.) in diameter. Since the average defect size for the glass slides, a_{i1}, is about 23 μm (0.0009 in.), Eq 15 can be used to calculate a value of 42 μm (0.0017in.) for a_{i2} for the cylinders. The distribution function (Eq 13) is then used to calculate the cumulative probability of failure versus defect size (Fig. 4). Now, for any probability of failure a stress rupture curve can be calculated for *any* environment if a crack growth rate curve exists for that environment. For example, for $P = 1\%$ (the weakest of 100 cylinders), the defect size is 107 μm (0.0042 in.). This defect size is characteristic of the abraded soda lime glass cylinders and is independent of environment. By using the constants for the crack growth rate curve for soda lime glass in water (Fig. 1), the average stress rupture curve for the cylinders (Eq 8) and the stress rupture curve for the weakest of 100 cylinders (tested in water) can be predicted (Fig. 5).

Conclusions

For an initial critical defect size, a_i, a stress rupture curve can be produced by integrating the analytical expression for a crack growth rate curve. For two

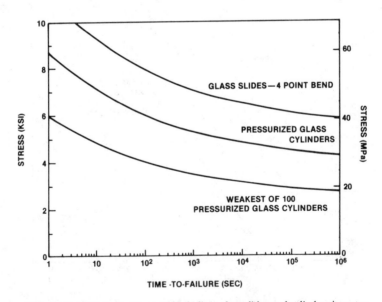

FIG. 5—*Stress rupture curves for soda lime glass slides and cylinders in water.*

different values of a_i, two stress rupture curves can be generated to represent scatter in rupture data from two specimens of the same size or average rupture curves for specimens with two different sizes. This scatter or size effect should be treated along lines of constant initial stress intensity factor. These lines have a slope of -0.5 on the log stress-log t_f curve. Along these lines the scatter or size effect in stress or life is independent of rupture life. An initial defect size can be inferred for a data point that falls on the rupture curve. Then, a size distribution of critical defects can be predicted for a set of rupture data. The Weibull distribution can be used to display the data in terms of $1/\sqrt{a_i}$. The size effect can be treated where the square root of the average size of the distribution of critical defects increases with effective area. Thus, critical defect size distributions in structures can be characterized so that allowable material defect sizes can be based on required component life.

Acknowledgment

The author would like to thank Curt Johnson for useful discussions concerning the analysis and John Ritter for supplying the detailed stress rupture data for soda lime glass.

References

[1] Barnett, R. L. and McGuire, R. L., "Statistical Approach to Analysis and Design of Ceramic Structures," *American Ceramic Society Bulletin*, Vol. 45, No. 6, 1966, pp. 595-602.

[2] Davies, D. G. S., "The Statistical Approach to Engineering Design in Ceramics," *Proceedings of the British Ceramic Society*, Vol. 22, 1973, pp. 429-452.

[3] Stanley, P., Fessler, H., and Sivill, A. D., "An Engineer's Approach to the Prediction of Failure Probability of Brittle Components," *Proceedings of the British Ceramic Society*, Vol. 22, 1973, pp. 453-487.

[4] Lenoe, E. M., "Probability-Based Design and Analysis—The Reliability Problem," *Ceramics for High Performance Applications*, Brook Hill, Chestnut Hill, Mass., 1974, pp. 123-145.

[5] Dukes, W. H., "Brittle Materials: A Design Challenge," *Mechanical Engineering*, November 1975, pp. 42-47.

[6] Paluszny, A. and Wu, W., "Probabilistic Aspects of Designing with Ceramics," *Journal of Engineering for Power*, Vol. 44, No. 4, 1977, pp. 617-630.

[7] Trantina, G. G. and deLorenzi, H. G., "Design Methodology for Ceramic Structures," *Journal of Engineering Power*, Vol. 99, No. 4, 1977, pp. 559-566.

[8] Ritter, J. E., Jr., "Engineering Design and Fatigue Failure of Brittle Materials," *Fracture Mechanics of Ceramics*, Vol. 4, R. C. Bradt, D. P. H. Hasselman, F. F. Lange, and A. G. Evans, Eds., Plenum, New York, 1978, pp. 667-686.

[9] Charles, R. J., "Static Fatigue in Glass," *Journal of Applied Physics*, Vol. 29, No. 11, November 1958, pp. 1549-1560.

[10] Mould, R. E. and Southwick, R. D., "Strength and Static Fatigue of Abraded Glass Under Controlled Ambient Conditions: II, Effect of Various Abrasions and the Universal Fatigue Curve," *Journal of the American Ceramic Society*, Vol. 42, No. 12, 1959, pp. 582-592.

[11] Wiederhorn, S. M. and Bolz, L. H., "Stress Corrosion and Static Fatigue of Glass," *Journal of the American Ceramic Society*, Vol. 53, No. 10, 1970, pp. 543-548.

[12] Ritter, J. E., Jr., and Sherburne, C. L., "Dynamic and Static Fatigue of Silicate Glasses," *Journal of the American Ceramic Society*, Vol. 54, No. 12, 1971, pp. 601-605.

[13] Ritter, J. E., Jr., and Meisel, J. A., "Strength and Failure Predictions for Glass and Ceramics," *Journal of the American Ceramics Society*, Vol. 59, Nos. 11-12, 1976, pp. 478-481.

[14] Trantina, G. G. and Johnson, C. A., "Probabilistic Defect Size Analysis Using Fatigue and Cyclic Crack Growth Rate Data," *Probabilistic Fracture Mechanics and Fatigue Methods*, *ASTM STP 798*, American Society for Testing and Materials, Philadelphia, 1983, pp. 67-78.

[15] Weibull, W., "A Statistical Theory of the Strength of Materials," *Proceedings of the Royal Swedish Institute of Engineering Research*, No. 151, 1939.

[16] Wiederhorn, S. M. and Johnson, H., "Effect of Electrolyte pH on Crack Propagation in Glass," *Journal of the American Ceramic Society*, Vol. 56, No. 4, 1973, pp. 192-197.

John E. Ritter, Jr.,[1] Karl Jakus,[1] and Robert C. Babinski[1]

Effect of Temperature and Humidity on Delayed Failure of Optical Glass Fibers

REFERENCE: Ritter, J. E., Jr., Jakus, K., and Babinski, R. C., "**Effect of Temperature and Humidity on Delayed Failure of Optical Glass Fibers,**" *Methods for Assessing the Structural Reliability of Brittle Materials, ASTM STP 844*, S. W. Freiman and C. M. Hudson, Eds., American Society for Testing and Materials, Philadelphia, 1984, pp. 131–141.

ABSTRACT: Existing and newly obtained optical glass fiber fatigue data over a wide temperature and humidity range were analyzed using the stress corrosion model of crack growth. Values obtained for the order of the reaction (humidity exponent), activation energy, and stress constant were compared with those obtained directly from crack growth studies. It was found that the data from different sources did not correlate well with each other and that none of the sets of data agreed with the stress corrosion model. It is believed that the polymer coatings on the fibers may play a complex role in determining the availability of moisture at the glass surface or that different fatigue mechanisms operate at different stress levels. Considering the uncertainty and ambiguity associated with modeling optical glass fiber fatigue data, an empirical approach to lifetime predictions is recommended in which a function that "best fits" the data obtained under service temperatures and humidities is used for prediction. It must be pointed out, however, that extrapolating outside the data range may result in inaccuracies in the predictions.

KEY WORDS: optical glass fibers, glass fibers, optical wave guides, fatigue, temperature effects, humidity effects, structural reliability, brittle materials

Mechanical reliability is a primary concern in the commercial use of optical glass fibers. These fibers are manufactured under carefully controlled processes that result in very high, short-term short length strengths (over 7000 MPa). Unfortunately, these fibers exhibit strength degradation that leads to delayed failure (commonly known as static fatigue) when put under moderate stress in a moist environment. For practical design purposes it is important to characterize this delayed failure process so that service lifetimes can be predicted from relatively short-term laboratory test data.

[1]Professors and research assistant, respectively, Mechanical Engineering Department, University of Massachusetts, Amherst, Mass.

Fracture mechanics [1–3] offers a framework within which lifetime predictions can be made. Unfortunately, long-term failure predictions are quite sensitive to the assumed crack growth model, and it is almost impossible to distinguish between fits of diverse expressions for crack growth over the normal data range for a given environment [4,5]. Thus, to model lifetime predictions successfully, a theory must also be able to account for the temperature and humidity dependence of the data. The purpose of this paper is to analyze existing and newly obtained fatigue strength data obtained over a wide range of temperature and relative humidity using the stress corrosion model of crack growth that has been applied successfully to other glasses [6–8].

Theory

It is generally believed that delayed failure is caused by the stress-enhanced growth of a preexisting flaw to dimensions critical for catastrophic failure. The theory that has demonstrated the best capability for modeling stress corrosion crack growth in glass systems is that due originally to Hillig and Charles [9] and later put into fracture mechanics terms by Wiederhorn [6]. These researchers assumed that crack velocity is governed by the rate of chemical reaction at the crack tip and, thus, that the crack velocity, v, is related to the stress intensity factor, K_I, by

$$v = ax_0^f \exp\left(\frac{-E^*}{RT}\right) \exp\left(\frac{bK_I}{RT}\right) \qquad (1)$$

where x_0 is the partial pressure of water, f is the order of the chemical reaction, E^* is an empirical measurement of the zero stress activation energy of the corrosion reaction, R is the gas constant, and a and b are constants.

Given a specific stress, temperature, and humidity history, Eq 1 can be integrated in principle to yield the lifetime of a particular glass fiber. For data analysis the two most important stress conditions are the constant applied stress (static fatigue) and the constant stressing rate (dynamic fatigue). Under constant stress, σ_a, Eq 1 gives [4]

$$t_f = \frac{2K_{Ic}}{AY^2 n} \left(\frac{1}{S_i \sigma_a} + \frac{1}{n\sigma_a^2}\right) \exp\left(-n\frac{\sigma_a}{S_i}\right) \qquad (2)$$

where $A = ax_0^f \exp(-E^*/RT)$; $n = bK_{Ic}/RT$; S_i is the inert strength, that is, the strength determined in an environment in which no subcritical crack growth occurs; and Y is a flaw shape factor.

Unfortunately, Eq 1 cannot be integrated in a closed form for constant stressing rate, $\dot{\sigma}$, and a numerical solution must be obtained from [4]

$$\int_{(K_{Ic}\dot{\sigma}/S_i)}^{(K_{Ic}\dot{\sigma}/\sigma_f)} \left(\frac{K_1}{t}\right) d\frac{K_1}{t} = \frac{Y^{22}A}{2}\int_0^{t_f} \exp(nK_1)dt \qquad (3)$$

where σ_f is the fracture strength.

The fatigue constants, A and n, can be determined as a function of temperature and humidity by regression analysis of constant stress or stressing rate data obtained at various temperatures and humidities using Eqs 2 and 3, respectively. Computer search techniques to perform the required nonlinear regressions of Eqs 2 and 3 have been previously published [4]. By analyzing the resulting temperature and partial pressure dependence of A and n, one can determine the stress-free activation energy E^*, the order of the reaction f, and the constants a and b.

Experimental Data

There are two sets of optical glass fiber fatigue data in the literature that were taken at varying temperatures and humidities. Figure 1 shows a typical plot from the constant stress rate (dynamic fatigue) data of Sakaguchi and Kimura [10], who varied both the temperature (20, 40, and 60°C) and the humidity (30, 40, 60, and 90% relative humidity). Figure 2 shows the plot of constant stress (static fatigue) data obtained by Chandan and Kalish [11] in water at various temperatures (40 to 90°C). It is apparent that Chandan and Kalish's data are substantially different from the data of Sakaguchi and Kimura, since they show that the fatigue mechanism at low stresses is different from that at high stresses.

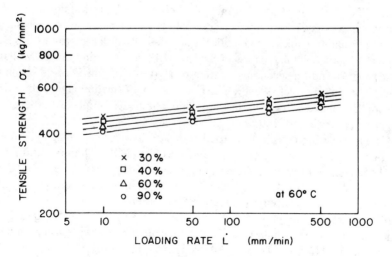

FIG. 1—*Example of constant stressing rate data of Sakaguchi and Kimura* [10].

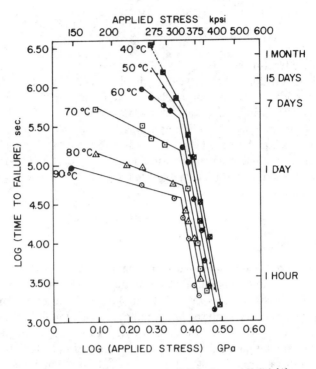

FIG. 2—*Constant stress data of Chandan and Kalish* [1].

To augment the two preceding sets of optical glass fiber data, additional constant stress data were obtained by the authors of this paper at varying temperatures (25, 42, and 65°C) and humidities (9 to 93% relative humidity). Figure 3 shows the results of these tests; the optical fiber in this study was a borosilicate clad, germainium-silicate core optical fiber made by ITT Electro-Optical Div., with a primary polymer coating of Sylgard 184 (Dow Corning Corp., Midland, Mich. 48690) and a secondary coating of Hytrel (E. I. du Pont de Nemours, Inc., Wilmington, Del. 19898). The tests were performed in an environmental chamber using dead-weight loading, and about 20 samples of 20-cm gage length were tested per condition.

The inert strength of the fiber used in this study was determined in dry nitrogen gas at 23°C, and the median value was 5946 MPa. Chandan and Kalish [11] determined their inert strength in dry carbon dioxide (CO_2) at -40°C and obtained a median value of 5820 MPa. Sakaguchi and Kimura [10] did not determine an inert strength for their fiber; therefore, the strength (13 390 MPa) obtained by Proctor et al [12] for fused silica fibers tested in liquid nitrogen was used in the analysis of their data. Duncan et al [13] also obtained a similar value (13 GPa) for the short-length liquid nitrogen strength of silica glass optical fibers. The exact values of "inert" strengths may be debatable,

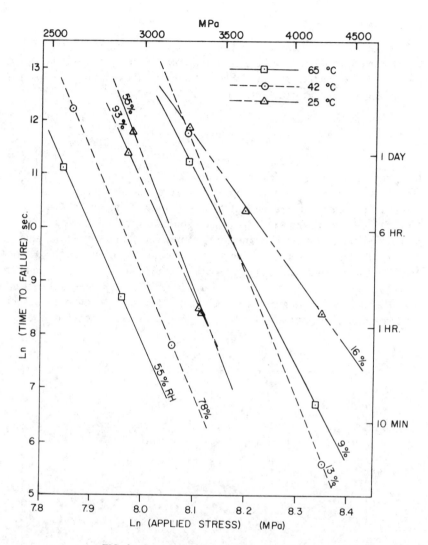

FIG. 3—*Constant stress data obtained in this study.*

but since inert strength does not play a primary role in the analysis of temperature and humidity dependence of fatigue parameters, these values were considered to be adequate for the present purposes.

The glass compositions in these three studies were not the same, which could
result in different fatigue susceptibilities. The purpose of this study, however,
was to compare the results to predictions by stress corrosion theories and not to
each other. Appropriate fatigue models must be developed before the fatigue
differences among various compositions may be meaningfully discussed.

Results and Discussion

The data of Sakaguchi and Kimura [10] and Chandan and Kalish [11] as well as the data obtained by the present authors were analyzed to obtain values for A and n at each of the available temperature and humidity conditions. Taking logarithms of the definition of A

$$\ln A = \ln a + f \ln x_0 - \frac{E^*}{RT} \qquad (4)$$

the order of the stress corrosion reaction, that is, the partial pressure exponent, can be determined as the slope of $\ln A$ versus $\ln x_0$ at constant temperature. As an example, Fig. 4 shows a plot based on the data of Sakaguchi and Kimura. It is evident that the slopes vary significantly with temperature (-0.8 to 0.5), indicating a possible change in the stress corrosion reaction. A value of E^* can be calculated from the slope of an Arrhenius plot of $\ln A$ versus $1/T$ at constant partial pressure; however, as can be seen from Fig. 4, the slope of A versus T depends on partial pressure for this set of data. Figure 5 is derived from the data in Fig. 4, and it is evident that different slopes are obtained for different partial pressures, effectively yielding a different value for the ac-

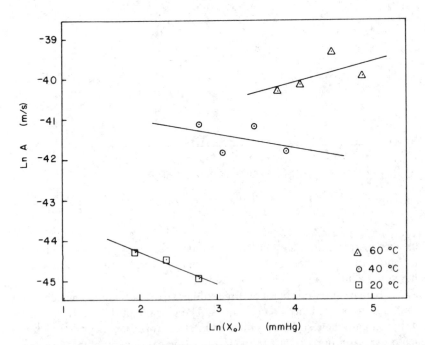

FIG. 4—*Dependence of fatigue parameter* A *on partial pressure of water, based on data of Sakaguchi and Kimura* [10].

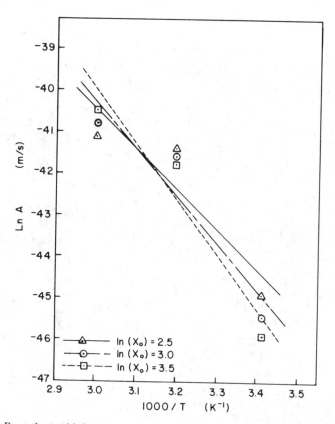

FIG. 5—*Dependence of fatigue parameter* A *on temperature, based on data of Sakaguchi and Kimura* [10].

tivation energy at each humidity level. Similarly, the constant n was found to depend not only on temperature, as expected from its definition, but also on partial pressure (see Fig. 6).

Since the value of b is determined from the slopes of the curves in Fig. 6, values of b varied from 0.43 to 0.57 $m^{5/2}/mol$. (One may note that the common intersection of the three regression lines on Fig. 6 is coincidental and, hence, bears no significance for the results.) Kalish and Tariyal [14] also found that the exponent, n, for fused silica optical fibers depended on humidity, contrary to stress corrosion theory. Although the supposedly environmentally indepen-dent constants, f, E^*, and b, showed a dependence on temperature and hu-midity with all of the data sets, values for these constants were calculated, and Table 1 shows the results of these calculations with the range of values ob-tained for each constant. From this table one can see not only the ambiguity of the constants that were obtained but also the inconsistencies between the dif-ferent sets of data. It should be noted that Chandan and Kalish's data were

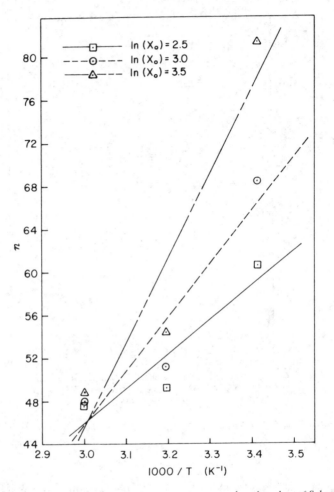

FIG. 6—*Dependence of fatigue parameter* n *on temperature, based on data of Sakaguchi and Kimura* [10].

separated into low- and high-stress regions for purposes of analysis. Also, since all their tests were done in water, there was no variability due to different humidity levels in determining the constants.

The stress corrosion constants in Table 1 can be compared with those obtained by Wiederhorn and Boltz [7], who measured crack growth for glasses in water over a temperature range of 2 to 90°C and as a function of relative humidity at 23°C. They found that for fused silica $E^* = 139$ kJ/mol and $b = 0.216$ m$^{5/2}$/mol, and for borosilicate glass $E^* = 129$ kJ/mol and $b = 0.200$ m$^{5/2}$/mol. In addition, they found that f was approximately 1, corresponding to the order expected for a silicon-oxygen bond breaking reaction, for

TABLE 1—*Summary of the stress corrosion constants based on fatigue data of optical glass fibers.*

E^*, kJ/mol	b, m$^{5/2}$/mol	f	Reference	Composition
109 to 231	0.318 to 0.488	−0.5 to 0.9	this study	borosilicate clad germanium-silicate core
55 to 92	0.435 to 0.572	−0.8 to 0.5	10	fused silica
29	−0.101	. . .	11	(high stresses) } fused silica
378	1.05	. . .	11	(low stresses) }

all humidity levels except at very low partial pressures. It is evident that agreement between the data for optical glass fibers and simple glass systems is not good.

Because of the poor correlation between the experimental constants and the stress corrosion model and the variability of these constants on temperature and humidity, calculations were made to minimize these environmental effects by analyzing the data with $f = 1$ and averaging the n values over humidity at each temperature. Table 2 summarizes these calculations, and, as before, the agreement between these crack propagation constants and those obtained directly from crack velocity measurements [7] is not good; furthermore, some of the values obtained for these constants are physically unacceptable (negative activation energy).

In summary, the excessive range of activation energies and the nonunity, even negative values, for the partial pressure exponent seem to indicate that the simple stress corrosion model is not adequate for representing fatigue data of optical glass fibers. The general lack of correlation between the fatigue data and the model is exhibited by the dependence on temperature and humidity of the supposedly environmentally independent constants, f, E^*, and b. Unfortunately, analysis with alternative crack velocity models [15-17] indicates that the problems encountered with the stress corrosion model as represented by Eq 1 recur in a similar fashion with the other models as well, that is, constants are not environmentally independent. It is possible that the polymer coating plays an active role in determining the accessibility of water at the crack tip, hence requiring a more complex expression for the effects of humidity or temperature, or both. It is also possible, as suggested by the data of Chandan and Kalish [11], that different mechanisms might be operable at high and low

TABLE 2—*Summary of the stress corrosion constants based on fixing f equal to 1 and averaging n values over humidity at each temperature.*

E^*, kJ/mol	b, m$^{5/2}$/mol	f	Reference
−0.291	0.124	1	this study
55	0.200	1	10

stresses. Furthermore, the assumption of preexisting flaws should be reexamined, considering that the strength-controlling flaw in optical glass fibers can be of atomic dimensions; hence, the concept of flaw initiation may be more appropriate. To resolve this current uncertainty associated with the applicability of the present models to fatigue data of optical glass fibers, more extensive fatigue data, as a function of temperature and humidity with a simple glass system, are needed to uncouple the possible role that the coating plays in the fatigue process. Until a better understanding of the fatigue of optical glass fibers is obtained, an empirical approach to design is the most prudent, in which data are obtained under temperature and humidity conditions representative of service, and predictions and extrapolations from these data are made with extreme caution using a function "best fit" to the data (power law, exponential, and so on). Kalish and Tariyal [14] demonstrated the utility of such an empirical approach by predicting static fatigue results from dynamic fatigue data obtained at the same temperature and humidity. It must be pointed out, however, that extrapolating outside the data range may result in substantial inaccuracies in predictions.

Acknowledgment

This research was funded by the U.S. Office of Naval Research under Contract N00014-78-C-0836.

References

[1] Ritter, J. E., Jr., "Engineering Design and Fatigue Failure of Brittle Materials," in *Fracture Mechanics of Ceramics*, Vol. 4, R. C. Bradt, D. P. Hasselman, and F. F. Lange, Eds., Plenum, New York, 1978, pp. 667-686.
[2] Freiman, S. W., "Fracture Mechanics of Glass," in *Glass Science and Technology. Vol. 5, Elasticity and Strength in Glasses*, D. R. Uhlmann and N. J. Kreidl, Eds., Academic Press, New York, 1980, pp. 21-78.
[3] Ritter, J. E., Jr., "Assessment of Reliability of Ceramic Materials," *Fracture Mechanics of Ceramics*, Vol. 5, R. C. Bradt, A. G. Evans, D. P. H. Hasselman, and F. F. Lange, Eds., Plenum Press, New York, 1983, pp. 227-251.
[4] Jakus, K., Ritter, J. E., Jr., and Sullivan, J. M., "Dependency of Fatigue Predictions on the Form of the Crack Velocity Equation," *Journal of the American Ceramic Society*, Vol. 64, 1981, pp. 372-374.
[5] Wiederhorn, S. M., in *Fracture 1977*, University of Waterloo Press, Waterloo, Ontario, Canada, 1977, pp. 893-901.
[6] Wiederhorn, S. M., "Influence of Water Vapor on Crack Propagation in Soda-Lime Glass," *Journal of the American Ceramic Society*, Vol. 50, No. 8, 1967, pp. 407-414.
[7] Wiederhorn, S. M. and Boltz, L. H., "Stress Corrosion and Static Fatigue of Glass," *Journal of the American Ceramic Society*, Vol. 53, 1970, pp. 543-548.
[8] Wiederhorn, S. M. and Johnson, H., "Effect of Electrolyte pH on Crack Propagation in Glass," *Journal of the American Ceramic Society*, Vol. 56, 1973, pp. 192-197.
[9] Charles, R. J. and Hillig, W. B., "The Kinetics of Glass Failure by Stress Corrosion," in *Symposium on Mechanical Strength of Glass and Ways of Improving It*, Florence, Italy, 25-29 September, 1961. Union Scientifique Continentale du Verre, Charleroi, Belgium, 1962, pp. 511-527.
[10] Sakaguchi, S. and Kimura, T., "Influence of Temperature and Humidity on Dynamic Fa-

tigue of Optical Fibers," *Journal of the American Ceramic Society*, Vol. 64, No. 5, 1981, pp. 259-262.

[*11*] Chandan, H. C. and Kalish, D., "Temperature Dependence of Static Fatigue on Optical Fatigue of Optical Fibers," *Journal of the American Ceramic Society*, Vol. 65, No. 3, 1982, pp. 171-173.

[*12*] Proctor, B. A., Whitney, I., and Johnson, J. W., "The Strength of Fused Silica," *Proceedings of the Royal Society A*, Vol. 297, 1966, pp. 534-557.

[*13*] Duncan, W. J., France, P. W., and Craig, S. P., "The Effect of Environment on the Strength of Optical Fiber," paper presented at NATO Workshop on Strength of Glass, Aramcao de Pera, Portugal, March 1983.

[*14*] Kalish, D. and Tariyal, B. K., "Static and Dynamic Fatigue of a Polymer-Coated Fused Silica Optical Fiber," *Journal of the American Ceramic Society*, Vol. 61, Nos. 11-12, 1978, pp. 518-523.

[*15*] Evans, A. G. and Wiederhorn, S. M., "Proof Testing of Ceramic Materials—An Analytical Basis for Failure Prediction," *International Journal of Fracture*, Vol. 10, No. 3, 1974, pp. 379-392.

[*16*] Lawn, B. R., "An Atomistic Model of Kinetic Crack Growth in Brittle Solids," *Journal of Materials Science*, Vol. 10, 1975, pp. 469-480.

[*17*] Brown, S. D., "Multibarrier Kinetics of Brittle Fracture: I, Stress Dependence of the Subcritical Crack Velocity," *Journal of the American Ceramic Society*, Vol. 62, Nos. 9-10, 1979, pp. 515-524.

DISCUSSION

G. D. Quinn[1] (written discussion)—There are many instances documented in the literature in which crack velocity experiments have failed to correlate with static fatigue experiments. Would you care to comment on this?

J. E. Ritter, Jr., K. Jakus, and R. C. Babinski (authors' closure)—Glass is one of the more exemplary materials to use in modeling fracture mechanics theory. At a given temperature and humidity, static fatigue can be predicted by fracture mechanics from dynamic fatigue data. The difficulty with optical glass fibers is that the effects of temperature and humidity on fatigue are not well understood. The reasons for the lack of correlation between static and dynamic fatigue data for some other materials are likely to be quite different from the reason optical glass fibers are not readily modeled by simple stress-corrosion reactions.

[1]U.S. Army-AMMRC, Watertown, Mass. 02172.

Timothy P. Dabbs,[1] *Carolyn J. Fairbanks,*[2] *and*
Brian R. Lawn[2]

Subthreshold Indentation Flaws in the Study of Fatigue Properties of Ultrahigh-Strength Glass

REFERENCE: Dabbs, T. P., Fairbanks, C. J., and Lawn, B. R., "Subthreshold Indentation Flaws in the Study of Fatigue Properties of Ultrahigh-Strength Glass," *Methods for Assessing the Structural Reliability of Brittle Materials, ASTM STP 844,* S. W. Freiman and C. M. Hudson, Eds., American Society for Testing and Materials, Philadelphia, 1984, pp. 142-153.

ABSTRACT: The rate-dependent characteristics of subthreshold indentation flaws in glass are surveyed. In the first part, the kinetics of radial crack initiation within the indentation field are described. It is shown that an incubation time must be exceeded in the contact process for a critical crack nucleus to develop. This incubation time decreases as the contact load and the water content in the environment increase. Even if incubation is not achieved *during* the contact, delayed pop-in may occur *after* the contact due to the action of residual stresses. Scanning electron microscopic evidence shows that the radial cracks initiate from precursor shear faults within the deformation zone. In the second part of the presentation, the fatigue properties of specimens with indentation flaws on either side of the threshold are discussed. The subthreshold flaws differ significantly from their postthreshold counterparts in these properties: the applied stresses at failure are higher, the susceptibility to water is stronger, and the scatter in individual data points is wider. These features are discussed in relation to the preceding crack-initiation kinetics. Finally, the implications of the results concerning design criteria for optical fibers are considered.

KEY WORDS: brittle materials, crack initiation, fatigue, glass, indentation flaw, optical fibers, radial crack, shear fault, strength, structural reliability

In an earlier paper in this volume [1] a strong argument was presented for the use of indentation flaws in the investigation of intrinsic fracture parame-

[1]Graduate student, School of Physics, University of New South Wales, New South Wales, Australia.
[2]Guest student and physicist, respectively, Center for Materials Science, National Bureau of Standards, Washington, D.C. 20234.

ters and flaw characteristics. One major advantage cited was the facility to control the scale of the flaw, by means of the contact load, in order to check for size effects in macroscopically determined crack growth laws. Such size effects would be apparent as deviations from the proposed "universal" plotting schemes. In view of the growing concern that microscopic flaws might differ in some significant respects from true cracks [2,3], particularly in the region of ultrahigh strengths (that is, approaching the theoretical limit), there would appear to be a need for systematic studies of flaw response at a fundamental level.

Accordingly, we survey here the results of some recent Vickers indentation studies on glass [4-6]. The original driving force for this work came from a pilot study on optical fibers [7] in which it was found that the strengths of specimens with indentations abruptly increased as the contact load diminished below some threshold. This threshold corresponded to a critical flaw configuration for crack *initiation*; the "pop-in" of radial cracks from the impression corners appeared to require the development of a sufficiently large precursor deformation zone [8,9]. Nevertheless, even though the subthreshold indentations were free of visible cracks, they still provided preferred sites for failure. It was therefore clear that the *mechanics*, if not the *mechanisms*, of failure were different at the subthreshold level. Thus, insofar as indentation flaws could be expected to simulate the broad features of certain natural flaw types in glass fibers, the conclusion could be drawn that extrapolations from macroscopic to microscopic domains, as is implicit in most (statistically based) strength analyses, may well be unjustified. Reported observations that the apparent crack velocity exponent evaluated from optical fiber fatigue data is distinctly lower (by about a factor of two [10-12]) than the corresponding parameter obtained from large-scale fracture specimens reinforce this conviction.

Our presentation will be made in two major parts. First, we shall summarize the results of some recent observations made on the kinetics of radial crack initiation in soda lime glass. It will be seen, from scanning electron micrographs of the indentations, that shear faults within the deformation zone act as the initiation precursors for the radial cracks. The threshold for pop-in is found to be highly sensitive to the contact period and to the presence of water. Fatigue data on the specimens indented at different loads, embracing the threshold, will then be presented. These data are interpreted in terms of initiation kinetics. The relevance of the interpretations to the mechanical behavior of optical fibers will remain an underlying motivation for the study.

Initiation Kinetics of Indentation Cracks

Microscopy of Flaws

In this section we discuss what we know about the nature of indentation flaws near the threshold load, as revealed by direct microscopical observations. The broader features of the typical deformation/fracture pattern, shown schematically in Fig. 1 for the Vickers indentation geometry, are readily discerned by conventional optical techniques; the scale of the flaw at radial crack pop-in is generally on the order 10 μm for glass. It is this amenability to optical observation that has led to the relatively advanced state of our understanding of flaw response in the postthreshold region [1].

To understand the response in the subthreshold region, however, it becomes necessary to focus attention on events *within* rather than *without* the deformation zone. For this purpose one must go to higher resolution techniques, such as scanning electron microscopy. Figures 2 and 3 are appropriate photomicrographs of indentations in soda lime glass. In the first of these figures we see two simple top views of the same, postthreshold indentation before and after etching in dilute hydrofluoric acid. The etching reveals the detailed structure of the hardness impression more clearly, although one must be careful here in drawing conclusions about the immediate postindentation configuration; it is apparent from a comparison of Fig. 2a and b that the acid treatment has caused additional pop-in events to occur at the indentation corners, under the action of the residual stress field [13]. In Fig. 3 we see both top and side views (obtained by positioning the Vickers pyramid across a preexisting hairline crack and then running this crack through the specimen [6]) of a subthreshold indentation. Again, there is evidence of considerable fine structure in the deformation zone, particularly in the subsurface section.

The features of greatest interest in the scanning electron micrographs are the well-defined "fault" traces on the free surface areas which define the irreversible deformation zone. These faults appear from their geometry to be shear activated, much the same as classic dislocation slip processes in crystalline materials, although they operate at stress levels close to the theoretical limit of cohesive strength. One may conceive of the faulting configurations in terms of an intermittent "punching" mode, whereby catastrophic slip takes place at periodic intervals along shear stress trajectories to accommodate the intense strains exerted by the penetrating indenter [6]. For further geometri-

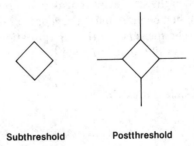

Subthreshold   Postthreshold

FIG. 1—*Schematic of Vickers indentation geometry. Cracks pop in spontaneously at the threshold to the well-developed radial configuration.*

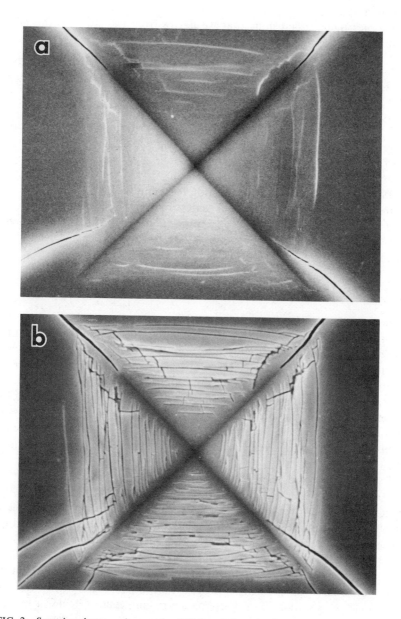

FIG. 2—*Scanning electron micrographs of Vickers indentation in soda lime glass. Top views: (a) unetched and (b) etched. The indentation load is 4 N; width of field is 40 μm. Radial cracks initiate from the shear fault structure in the hardness zone.*

cal details of the fault patterns, reference is made to the important papers by Hagan and Swain [14,15] on the subject.

The pertinent conclusions that we derive from the study of the micrographic evidence are as follows [6]:

1. The freshly created fault walls tend to recontact due to strongly compressive normal stresses within the contact zone, and thence to heal at the interface [6]. This healing is, nevertheless, incomplete since the faults remain susceptible to preferential etching.

2. Faults generated in different quadrants of the contact area can intersect to produce high stress intensifications (notwithstanding a certain ability for the faults to interpenetrate [15]), thus providing favorable sites for crack nucleation and growth.

3. The tendency for fault production to be intermittent (possibly because of the availability of suitable defect centers for the initial generation) leads to a certain variation in the geometrical disposition of the pattern from indentation to indentation (and even from quadrant to quadrant), thus giving rise to an element of scatter in the distribution of crack nuclei.

Kinetics of Crack Pop-In

It was indicated in the preceding subsection that optical techniques, although of insufficient resolution to observe the finer details of the precursor initiation processes, are usually adequate for observing the actual pop-in event itself. Here we describe the results of such observations made *in situ* through a microscope objective located immediately below the indentation site [6]. In our setup, a sinusoidal load-time pulse is applied to the Vickers indenter, and the time for radial pop-in, t_c, after the start of contact is recorded in relation to the indentation period, T. The functional interdependence of these two time variables then characterizes the kinetic response of the indentation flaws.

Typical results are shown in Fig. 4 for soda lime glass tested in air at a prescribed maximum load, P. It is immediately clear that the threshold condition is rate sensitive. At long contact periods pop-in occurs *during* the indentation cycle, specifically at a reproducible point $0.3\,P$ on the unload half cycle. At short contact periods, however, pop-in occurs *after* the contact, generally with extensive delays, $t_c \gg T$, and with large scatter. These results show that the driving force for initiation is greatest while the indenter is under load but is nevertheless far from insignificant in the residual field. One may interpret the transition from short-contact to long-contact behavior in terms of an *incubation* time for development of a critical nucleus from the shear fault configuration. This incubation time is found to diminish as the peak contact load or the water content of the environment increases [6], consistent with the general fatigue experience.

Observations of this type lead us to the following picture of initiation kinet-

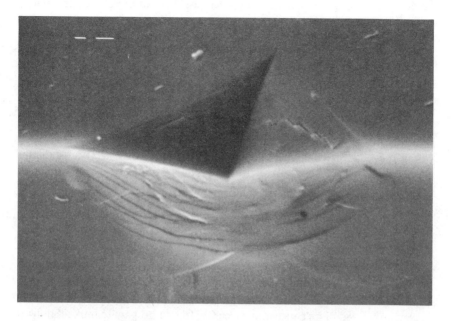

FIG. 3—*Scanning electron micrograph of Vickers indentations in soda lime glass. Simultaneous top and side view. The indentation load is 1 N; width of field is 25 μm. Shear faults are evident below the hardness impression.*

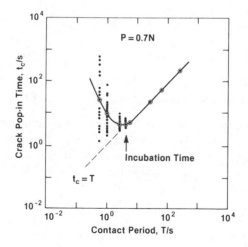

FIG. 4—*Time to fracture as a function of the contact period for soda lime glass in air. The open symbols denote median values at each prescribed contact period. The results indicate an incubation time for development of a critical crack nucleus.*

ics, based on the two-step process of precursor shear faulting and subsequent tensile crack pop-in [6]:

1. Newly formed shear faults become exposed to moisture in the environment, which penetrates into the interfaces and reduces the (already incomplete) cohesion. The penetration and decohesion mechanisms are both rate dependent, and therefore have the potential for controlling the kinetics. As this stage progresses, the stress intensifications at mutual fault intersection points increase, thereby building up the driving forces on the crack nuclei centers. It can be shown that the shear stress component responsible for activating the fault mechanisms reaches its maximum at the peak of indenter penetration but remains in force, albeit at reduced intensity, after completion of the cycle, consistent with the observations in Fig. 4.

2. The crack nuclei form close to the specimen surface, and are therefore similarly exposed to environmental moisture. Thus, unless the stress conditions are such that the nuclei propagate spontaneously to the fully developed radial configuration, the system will be subject to slow crack growth. The normal stresses on such nuclei remain compressive through most of the indentation cycle, becoming tensile only at the $0.3\,P$ unload point. If the incubation time has been exceeded by the time this point is reached the critical nucleus, released of its constraint, becomes free to propagate immediately; if not, the incubation process continues in the weakened, residual stress field, in which case the pop-in kinetics become subject to the variability in nucleation centers alluded to earlier.

According to this description, either of these two steps could control the rate at which the critical stress intensification develops at the initiation center. Further work is needed to establish the relative importance of the shear fault and tensile crack concepts in the initiation kinetics.

Fatigue of Glass with Postthreshold and Subthreshold Indentation Flaws

We now examine the results of dynamic fatigue studies of soda lime glass containing postthreshold [4] and subthreshold [5] flaws. These studies were carried out using annealed rod specimens of 4 to 5-mm diameter, the surfaces of which had been preetched to remove large handling flaws and had been coated with protective lacquer over all but a central region ≈ 3 mm wide. Vickers indentations were placed in these uncoated test areas at prescribed peak loads, and were examined after 30 min to determine whether radial cracks had or had not popped in. The specimens were then stressed at constant rates in four-point flexure in water, with the indentation oriented for maximum tension, and their fatigue strengths duly recorded. In each case the broken specimens were checked to confirm that the failures had indeed originated from the indentation site; those that had not were omitted from the data accumulation.

Figure 5 shows the results for both postthreshold and subthreshold flaws, plotted in accordance with the master map scheme advocated in Ref *1*. We recall from this earlier source that, for indentations with well-developed cracks, the dynamic fatigue response for any given material may be represented by the expression

$$\sigma_f P^{1/3} = (\lambda_P' \dot\sigma_a P)^{1/(n'+1)} \tag{1}$$

where P is the indentation load, $\dot\sigma_a$ is the stressing rate, σ_f is the strength, and n' and λ_P' are adjustable parameters. Both n' and λ_P' are load independent, so plotting in the coordinates of Fig. 5 should reduce all the data to a universal curve, regardless of the scale of the flaws. It is clear that there is a breakdown in this universality as one traverses the threshold into the low-load region:

1. The strengths are substantially higher, by a factor of three to four, consistent with the earlier study on fibers [7].

2. The scatter in data is wider, in correlation with the element of variability observed in the initiation kinetics (Fig. 4).

3. The susceptibility to fatigue is stronger; compare the apparent crack velocity exponents $n' = 9.0 \pm 0.8$ for subthreshold and $n' = 14.0 \pm 0.3$ for postthreshold.

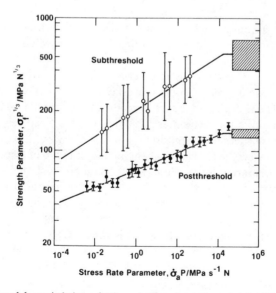

FIG. 5—*Reduced dynamic fatigue plot for soda lime glass rods containing Vickers indentation flaws tested in water. The shaded bands indicate inert strength levels; the error bars are standard deviations; load range for postthreshold data is 0.05 to 10 N (see Ref 1) and for subthreshold data, 0.15 to 0.25 N.*

Equation 1 is derived on the basis of specific assumptions concerning the flaw configuration, notably that the critical crack extension occurs in the far field of the initiation zone [1]. Therefore, it should not be altogether surprising that the subthreshold data plot onto a different curve in Fig. 5. The characteristics of the data in the two regions are, nevertheless, sufficiently diverse to suggest that we are dealing with two entirely different failure mechanisms, corresponding to a transition from propagation- to initiation-controlled instabilities. In this context it may be noted that the inert strength level for postthreshold flaws lies below the fatigue strength for subthreshold flaws over a large portion of the stressing rate range; thus, when radial cracks do initiate during a fatigue test, failure must occur spontaneously from the precursor shear fault configuration.

It is useful to replot the data in Fig. 5 in a way which brings out the scale effect more clearly. This is done in Fig. 6. To obtain this plot we have made use of the lifetime analogue of Eq 1 [1]; that is, substituting $\dot{\sigma}_a = \sigma_f / t_f$ and identifying σ_f with the constant applied stress σ_A, rearranged thus

$$t_f \sigma_A^{n'} = \lambda_P' / P^{(n'-2)/3} \tag{2}$$

as an explicit function of load. Each point in Fig. 6 accordingly represents a composite evaluation of $t_f \sigma_A^{n'}$ from all data at a given P, taking the two regions separately but using n' from the *postthreshold* curve fit in Fig. 5 as an appropriate fatigue exponent. The rationale for this choice of a common exponent is that, as mentioned before, the derivation of Eqs 1 and 2 is based strictly on the assumption that failure occurs from well-developed cracks. Thus, flaws which violate this assumption would be expected to show systematic departures from the baseline data curve, and this is indeed observed to be the case in Fig. 6. The solid line in this figure is equivalent to the prediction we would make using macroscopic crack laws (with due account of residual stress terms [1]); it is seen that such a prediction, extrapolated into the subthreshold domain, underestimates the lifetime at a fixed level of applied stress by several orders of magnitude.

It is useful to view the threshold phenomenon in Fig. 6 in terms of flaw size rather than indentation load. We have, accordingly, included a scale of the hardness impression a on the abscissa, using the hardness relation $H = P/2a^2 = 5.5$ GPa for Vickers indentations on glass as the basis for conversion. In the present case the threshold occurs in the range $a = 1$ to 10 μm, although it must be remembered that this range can scale up or down, depending on the kinetic history of the flaw. The overlap between the subthreshold and postthreshold data is attributable to the variability in the shear fault configurations discussed earlier. Thus, it is conceivable that a particular flaw could be substantially smaller in its characteristic dimension than many of its possible competitors in a given specimen, and yet still control the strength properties by virtue of some uniquely favorable crack initiation conditions. Of

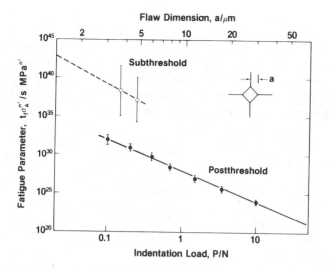

FIG. 6—*Fatigue data from Fig. 5, replotted to show load (or equivalent flaw size) dependence. The error bars are standard deviations.*

course, the potential always remains in such instances for delayed pop-in from some of these larger flaw centers, with consequent degradation in the overall lifetime response.

Implications Concerning Optical Fibers

We conclude this discourse by examining the implications of the results on the fatigue properties of ultrastrong silica fibers. The data presented here were obtained on rod rather than fiber specimens, and on soda lime rather than high silica glass. Although we have taken precautions to eliminate pre-existing handling flaws on our rod specimens, we can never be certain that the pristine surfaces of freshly drawn fibers would have responded in exactly the same way, and it is now well known that soda lime and high silica glasses differ significantly in their indentation behavior [16], the latter being regarded as anomalous in this respect. Nevertheless, it may be recalled that our very first observations of threshold effects in the strength behavior were on actual optical fibers [7], so we may feel justified in making certain predictions, if only of a qualitative nature.

One point that comes out clearly in the present work is the need to identify the flaw region which characterizes the operating stress range for a given component. Thus, if we wished to use the type of glass represented by the data in Figs. 5 and 6 as a structural material at ultrahigh strengths, it would hardly be efficient to design on the basis of extrapolations from the macroscopic crack region. (On the other hand, such extrapolations, in this case at least, would suffice in applications calling for extreme conservatism in lifetime predictions.) These comments would appear to reinforce the conviction held by many that, as far as practically possible, design evaluations should be made using data taken on surfaces which have the same finish as those to be placed in service and which embrace the required lifetime. This is particularly so when adopting empirical approaches to flaw characterization, especially those based on statistical distribution functions, where the underlying processes responsible for creating the flaws in the first place are disregarded.

The point just made about designing within the time range of the fatigue data bears further elaboration here, for, as we have seen, flaws can continue to evolve long after their inception. Accordingly, where it is necessary to make extended lifetime predictions from relatively short-term fatigue data, the possibility exists for premature failures. Any amount of fatigue testing of control samples, or proof testing of actual components, will count for nothing if surfaces with subthreshold flaws are subsequently exposed to service environments conducive to crack initiation. There is some evidence from the glass fiber literature [17,18] that small-scale *natural* flaws can indeed suffer abrupt increases in severity with prolonged exposure to water, although in these cases direct observations of the flaws themselves could not be made to establish the nature of the transition. There would, accordingly, seem to be a strong case for advocating more controlled studies of fatigue failure in optical fibers, using flaws introduced by indentation or other artificial means, so that the fundamental lifetime-controlling processes might ultimately be identified and studied in a more systematic manner.

Acknowledgments

The authors gratefully acknowledge many useful discussions with R. F. Cook. Funding was provided by the U.S. Office of Naval Research (Metallurgy and Ceramics Program).

References

[1]   Cook, R. F. and Lawn, B. R., in this publication, pp. 22-42.
[2]   Lawn, B. R., *Journal of the American Ceramic Society*, Vol. 66, 1983, pp. 85-91.
[3]   Jakus, K., Ritter, J. E., and Babinski, R. C., in this publication, pp. 131-141.
[4]   Dabbs, T. P., Lawn, B. R., and Kelly, P. L., *Physics and Chemistry of Glasses*, Vol. 23, 1982, pp. 58-66.
[5]   Dabbs, T. P. and Lawn, B. R., *Physics and Chemistry of Glasses*, Vol. 23, 1982, pp. 93-97.
[6]   Lawn, B. R., Dabbs, T. P., and Fairbanks, C. J., *Journal of Materials Science*, Vol. 18, 1983, pp. 2785-2797.
[7]   Dabbs, T. P., Marshall, D. B., and Lawn, B. R., *Journal of the American Ceramic Society*, Vol. 63, 1980, pp. 224-225.
[8]   Lawn, B. R. and Evans, A. G., *Journal of Materials Science*, Vol. 12, 1977, pp. 2195-2199.
[9]   Lawn, B. R. and Marshall, D. B., *Journal of the American Ceramic Society*, Vol. 62, 1979, pp. 347-350.
[10] Ritter, J. E. and Jakus, K., *Journal of the American Ceramic Society*, Vol. 60, 1977, p. 171.

[*11*] Kalish, D. and Tariyal, B. K., *Journal of the American Ceramic Society*, Vol. 61, 1978, pp. 518-523.

[*12*] Gulati, S. T., Helfinstine, J. D., Justice, B., McCartney, J. S., and Runyan, M. A., *American Ceramic Society Bulletin*, Vol. 58, 1979, pp. 1115-1117.

[*13*] Dabbs, T. P. and Lawn, B. R., *Journal of the American Ceramic Society*, Vol. 65, 1982, pp. C37-C38.

[*14*] Hagan, J. T. and Swain, M. V., *Journal of Physics D: Applied Physics*, Vol. 11, 1978, pp. 2091-2102.

[*15*] Hagan, J. T., *Journal of Materials Science*, Vol. 15, 1980, pp. 1417-1424.

[*16*] Arora, A., Marshall, D. B., Lawn, B. R., and Swain, M. V., *Journal of Non-Crystalline Solids*, Vol. 31, 1979, pp. 415-428.

[*17*] Chandan, H. C. and Kalish, D., *Journal of the American Ceramic Society*, Vol. 65, 1982, pp. 171-173.

[*18*] Gupta, P. K., in *Fracture Mechanics of Ceramics*, R. C. Bradt, A. G. Evans, D. P. H. Hasselman, and F. F. Lange, Eds., Plenum, New York, 1982, pp. 291-303.

Theo Fett[1] and Dietrich Munz[2]

Lifetime Prediction for Hot-Pressed Silicon Nitride at High Temperatures

REFERENCE: Fett, T. and Munz, D., **"Lifetime Prediction for Hot-Pressed Silicon Nitride at High Temperatures,"** *Methods for Assessing the Structural Reliability of Brittle Materials, ASTM STP 844,* S. W. Freiman and C. M. Hudson, Eds., American Society for Testing and Materials, Philadelphia, 1984, pp. 154–176.

ABSTRACT: Four-point bend specimens of hot-pressed silicon nitride were tested at 1000 and 1200°C at constant displacement rates and with constant applied load. At all temperatures the relationship between failure probability and time to failure could be predicted for the tests with constant load from the test performed at constant displacement rate, using the relations of linear elastic fracture mechanics. Regarding the test at 1200°C, the varying stresses due to creep have to be taken into account in the predictions.

KEY WORDS: silicon nitride, crack propagation, lifetime prediction, creep, scatter, structural reliability, brittle materials

The lifetime of statically loaded ceramic structural components is restricted by subcritical crack growth. Preexisting flaws may extend at increasing velocity up to critical sizes for spontaneous propagation, followed by failure of the loaded component.

There are two different fracture mechanical procedures to prevent this delayed failure within the planned time of operation. One of them is the proof test method. The theoretical and experimental foundations are reported in detail by Evans and Wiederhorn [1]. A second method is the lifetime prediction by measurement of crack growth parameters and preexisting flaw size distribution, based on statistical treatments of strength data. This well-known procedure has been applied successfully to several ceramic materials in case of linear elastic material behavior. Measurements on glass in water [2–4] have given good agreement between predicted and experimentally obtained lifetimes; the same is true for investigations on alumina in water [5].

[1]Research engineer, Kernforschungszentrum Karlsruhe, Arbeitsgruppe Zuverlässigkeit und Schadenskunde am Institut für Reaktorbauelemente, West Germany.
[2]Professor, Universität Karlsruhe, Institut für Zuverlässigkeit und Schadenskunde im Maschinenbau, West Germany.

Difficulties appear at high temperatures, where creep behavior dominates. Trantina [6] observed nonlinear load-displacement curves for hot-pressed silicon nitride at 1200°C; nevertheless, he applied the linear elastic relations. For the same material, Wiederhorn and Tighe [7] applied the proof test method and discussed the problem of different flaw size populations at room temperature and at elevated temperatures. Lange [8] considers an interrelationship between creep and crack growth.

In this paper the authors demonstrate a way to modify the usually employed formulas so that lifetime predictions can be executed under creep conditions.

Time to Failure in Case of Elastic Behavior

In the case of elastic behavior, lifetime predictions can be performed with elementary relationships. The time to failure under constant tensile stress, σ, can be derived from the definition of the stress intensity factor, K

$$K = \sigma Y \sqrt{a} \qquad (1)$$

where a is a flaw size parameter and Y depends on the flaw size and geometry of the specimen, and from the well-known power law of subcritical crack growth

$$v = \frac{da}{dt} = AK^N \qquad (2)$$

where A and N are parameters of the material. Combining Eq 1 and Eq 2 gives

$$t_f = \frac{2}{A\sigma^2 Y^2} \int_{K_i}^{K_c} \frac{dK}{K^{N-1}} = \frac{2}{A\sigma^2 Y^2(N-2)} (K_i^{2-N} - K_c^{2-N}) \qquad (3)$$

where $K_i = \sigma Y \sqrt{a_i}$ is the initial stress intensity factor, a_i the initial crack size, and K_c the fracture toughness. Y is assumed to be independent of crack size. For ceramic material, one obtains $N > 10$, and in the case of $K_i < 0.9 K_c$ we can neglect K_c^{2-N}. Then Eq 3 is simplified to read

$$t_f = B\sigma_c^{N-2}\sigma^{-N} \qquad (4)$$

with

$$B = \frac{2}{AY^2 (N-2)K_c^{N-2}}$$

and

$$\sigma_c = \frac{K_c}{Y\sqrt{a_i}}$$

the inert strength (strength without any crack propagation).

Lifetimes can be predicted by Eq 4 if the parameters N and B and the inert strength σ_c are known.

In the same manner, we can calculate the strength σ_B as a function of the stress rate $\dot\sigma$ and the inert strength σ_c if we replace dt in Eq 2 by $dt = d\sigma/\dot\sigma$. The result of this analysis is [9]

$$\sigma_B^{N+1} = B\sigma_c^{N-2}(N + 1)\dot\sigma \tag{5}$$

This relationship makes it possible to determine N and $B \cdot \sigma_c^{N-2}$, which is necessary for lifetime predictions. In a plot log σ_B versus log $\dot\sigma$ a straight line is expected with the slope $1/(N + 1)$ and an intercept of

$$(N + 1) \log \sigma_B = \log (B\sigma_c^{N-2}) + \log (N + 1)$$

at $\dot\sigma = 1$ (in convenient units).

Generally, the strength of ceramics is a distributed value and is given by the Weibull relationship

$$\ell n \, \ell n \, \frac{1}{1 - F} = m \, \ell n \, \frac{\sigma_c}{\sigma_{c0}} \tag{6}$$

where F = probability of failure and m, σ_{c0} are Weibull parameters.

If the inert strength σ_c can be described by a Weibull distribution, then also the strength σ_B in a test with constant $\dot\sigma$ and the lifetime t_f in a test with constant σ follow the Weibull distributions

$$\ell n \, \ell n \, \frac{1}{1 - F} = m' \, \ell n \, \frac{\sigma_B}{\sigma_{B0}} \tag{7a}$$

$$\ell n \, \ell n \, \frac{1}{1 - F} = m_t \, \ell n \, \frac{t_f}{t_{f0}} \tag{7b}$$

with

$$m' = m \, \frac{N + 1}{N - 2} \, ; \qquad m_t = \frac{m}{N - 2}$$

$$\sigma_{B0} = [B\sigma_{c0}^{N-2}(N + 1)\dot\sigma]^{1/N+1} \, ; \qquad t_{f0} = B\sigma_{c0}^{N-2}\sigma^{-N}$$

Determination of N Under Creep Conditions

Determination of Creep Exponent

At high temperatures the deformation behavior of ceramics is governed mainly by creep. In this case, the linear elastic equations (3 to 5) are no longer valid.

The creep behavior can be described by creep curves. These curves can be expressed by analytical formulas. One of the best known is

$$\epsilon = \epsilon_{el} + \epsilon' + \epsilon'' \tag{8a}$$

$$\epsilon = \frac{\sigma}{E} + \sigma^n C(1 - e^{-\alpha t}) + \sigma^n Bt \tag{8b}$$

where

ϵ_{el} = elastical strain,
ϵ' = transient creep strain,
ϵ'' = secondary creep strain,
E = Young's modulus, and
α, C, B, n = constants at constant temperature.

Equation 8b is only valid in case of constant stresses. For applications with changing stresses so-called hardening rules are applied: the time hardening rule or the more realistic strain hardening rule. In hardening formulation, an arbitrary given time dependency, $\sigma = f(t)$, is composed by piecewise constant stresses, and the accompanying strain increments are picked up from constant stress curves (Eq 8b). If strain rate $\dot{\epsilon}$ is expressed in the form $\dot{\epsilon} = f(\sigma, t)$, the time hardening formula is obtained. If, for the transient part, time is replaced by strain, the strain hardening formula $\dot{\epsilon} = f(\sigma, \epsilon')$ is obtained. In time hardening formulation it yields from Eq 8b

$$\dot{\epsilon} = \frac{\dot{\sigma}}{E} + \sigma^n \frac{d}{dt} [C(1 - e^{-\alpha t}) + Bt] = \frac{\dot{\sigma}}{E} + C\alpha\sigma^n e^{-\alpha t} + B\sigma^n \tag{8c}$$

Note that no term with $d/dt(\sigma^n)$ occurs!

Substitution of Eq 8b into Eq 8c delivers the strain hardening formula for variable stresses used by Pao and Marin [10]

$$\dot{\epsilon} = \frac{\dot{\sigma}}{E} + \alpha(C\sigma^n - \epsilon') + B\sigma^n \tag{8d}$$

From measurements on hot-pressed silicon nitride (HPSN) [11,12], it can be concluded that $1/\alpha$ is on the order of 10 h. The short-time creep behavior ($\alpha t \ll 1$ in Eq 8c) can be described by a constant transient creep rate

$$\dot{\epsilon} = \frac{\dot{\sigma}}{E} + D\sigma^n \qquad (9)$$

with $D = \alpha C + B$.

For tensile tests performed at a constant strain rate, $\dot{\epsilon}$, integration of Eq 9 leads to

$$\int_0^{\frac{\sigma}{\sigma_\infty}} \frac{d\left(\dfrac{\sigma}{\sigma_\infty}\right)}{1 - \left(\dfrac{\sigma}{\sigma_\infty}\right)^n} = \frac{\sigma_e}{\sigma_\infty} \qquad (10)$$

with the substitutions

$$\sigma_\infty = \left(\frac{\dot{\epsilon}}{D}\right)^{1/n}, \qquad \sigma_e = \dot{\epsilon}Et \qquad (11)$$

Equation 10 can be solved analytically if n is an integer. Unfortunately, solutions are given implicitly, if $n > 2$. For $n > 1.5$, a numerical representation can be given within $\pm 0.5\%$ accuracy by [5]

$$\frac{\sigma}{\sigma_\infty} = \left[\tanh\left(\frac{\sigma_e}{\sigma_\infty}\right)^p\right]^{1/p} \qquad (12a)$$

or

$$\sigma = \left(\frac{\dot{\epsilon}}{D}\right)^{1/n}\left[\tanh\left(\frac{D^{1/n}E}{\dot{\epsilon}^{1/n}}\epsilon\right)^p\right]^{1/P} \qquad (12b)$$

where $p = 0.207 + 0.3965n$.

Equation 12 is exact for $n = 2$. If Eq 9 were valid also for large strains, then for $t \rightarrow \infty$ or $\epsilon \rightarrow \infty$, the stress would approach σ_∞.

Equation 12b can be applied to dynamic bending tests. The bending moment M at constant deflection rate can be calculated by integrating the stress distribution over a beam with rectangular cross section.

$$M = 2b\int_0^{h/2} \sigma(x)x\,dx \qquad (13)$$

with

b = thickness of the beam,
h = height of the beam, and
x = distance from the neutral axis.

Two assumptions are made:

1. The strain and strain rate vary linearly with the distance from the neutral axis.
2. The stresses are symmetrically distributed in cross section.

This frequently applied assumption may not be fulfilled for all ceramic materials. It was found that creep under tension is distinctly higher than creep under compression, especially in steady-state creep range where tensile creep is accompanied by the formation of cavities and development and growth of cracks [13]. For transient creep, this effect could be small.

The solutions so obtained are depicted in Fig. 1 as

$$\frac{M}{M_\infty} = f\left(n, \frac{M_e}{M_\infty}\right)$$

where

$$M_e = \sigma_e^s \cdot W = \dot{\epsilon}^s E t W \qquad M_\infty = \frac{3n}{2n + 1}\sigma_\infty^s W \qquad (14)$$

and $W = bh^2/6$, σ_e^s, and σ_∞^s are the outer fiber stresses ($x = h/2$).[3] Figure 1 leads to a simple procedure of determining creep exponents, n, from the experimentally determined relationship between bending moment and deflection or time, respectively.

The slope in the originally linear part of such a curve should be determined and two straight lines with 90 and 95% of this slope drawn in the diagram. Intersections between the straight lines and curve give the values $M(90)$ and $M(95)$. The quotient $M(95)/M(90)$ is a characteristic value of the shape of each curve.

The relationship between $M(95)/M(90)$ and n, based on the computed curves of Fig. 1, is represented in Fig. 2. In addition, the quotient $M(90)/M_\infty$ is depicted there and also the procedure for determination of $M(90)$, $M(95)$.

The curves in Fig. 2 can be described by

$$\frac{1}{n} = a_0 + a_1 X + a_2 X^2 + a_3 X^3 + a_4 X^4 \qquad (15)$$

---

[3]From here on the superscript s to characterize the outer fiber is omitted.

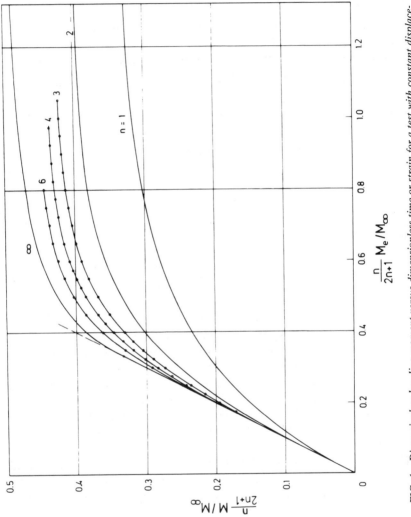

FIG. 1—*Dimensionless bending moment versus dimensionless time or strain for a test with constant displacement rate under creep conditions.*

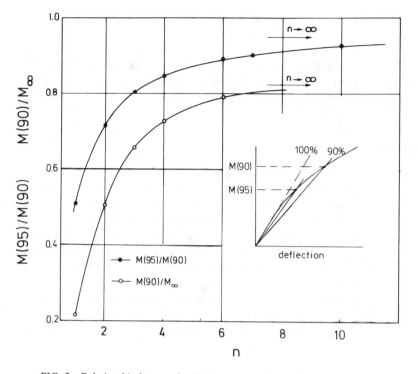

FIG. 2—*Relationship between bending moment ratios and creep exponent, n.*

with

$$X = 0.948 - M(95)/M(90); X \geq 0.015,$$
$$a_0 = 0.05858, a_1 = 1.7756, a_2 = 1.2182, \text{ and}$$
$$a_3 = -4.9715, a_4 = 9.3219;$$

and

$$\frac{M(90)}{M_\infty} = \frac{2n + 1}{3n}\left(b_0 + \frac{b_1}{n} + \frac{b_2}{n^2} + \frac{b_3}{n^3} + \frac{b_4}{n^4}\right) \qquad (16)$$

with

$$b_0 = 0.824,$$
$$b_1 = -0.1545,$$
$$b_2 = -3.1984,$$
$$b_3 = 4.653,$$
$$b_4 = -1.981, \text{ and}$$
$$1 \leq n \leq 20.$$

Thus n can be obtained from Eq 15 using $M(95)/M(90)$, and M_∞ can be obtained from Eq 16 using n and $M(90)$. From bending strength tests on

HPSN at 1200°C, an average creep exponent of $n \simeq 7$ was found [5]. In case of low bending stresses, values in the range $1 < n < 3$ are often reported in literature [11,12]. Arons and Tien [14] obtained values of $3.5 < n < 5.5$. At high stresses—that are necessarily reached in bending strength tests—distinctly higher values of n were found. Measurements of Kossowsky et al [11] gave at 1160°C $n \approx 50$ for $\sigma > 100 \ N/mm^2$.

Determination of Exponent N

In the absence of creep, the exponent N can be determined from Eq 5 where the outer fiber fracture stress, σ_B, is the elastically calculated stress σ_e at fracture and the outer fiber stress rate $\dot{\sigma}_e = \dot{\epsilon}E$. Because of the creep effects, $\dot{\sigma}$ is not constant in a test carried out at constant deformation rate, and σ_B is not obtained simply from M_B / W, where M_B = bending moment at failure. In the following, it is assumed that linear elastic fracture mechanics relations can still be applied to describe crack propagation. Creep only affects the global stress distribution in the specimen.

In the past, the assumption of linear elastic fracture mechanics very often was made in describing subcritical crack growth, for instance measured with a double torsion specimen [15,16]. Nevertheless, this assumption can be de-

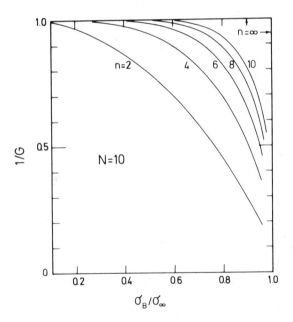

FIG. 3—*Stress rate correction factor for tests with constant displacement rate versus stress at failure.*

bated. Experimental results in connection with a detailed evaluation of the strain field at the crack tip can prove the validity of this assumption.

A detailed analysis [5] yields a relationship between the outer fiber stress at fracture σ_B and $\dot{\sigma}_e$, similar to Eq 5, with a correction factor $1/G$ dependent on N, n, σ_B

$$\sigma_B^{N+1} = \frac{\dot{\sigma}_e}{G\left(N, n, \dfrac{\sigma_B}{\sigma_\infty}\right)} B(N + 1)\sigma_c^{N-2} \qquad (17)$$

The true outer fiber fracture stress, σ_B, can be obtained from Eq 12a with $\sigma_e = E\dot{\epsilon}t$, because for the outer fiber a constant strain rate can be assumed. σ_∞ can be obtained from Eq 14 with M_∞ from Eq 16.

In Fig. 3 the calculated relationship between $1/G$ and σ_B/σ_∞ is shown for $N = 10$ with n as a parameter. In Fig. 4 the effect of N is depicted. For the determination of the exponent N from a log σ_B − log $(\dot{\sigma}_e/G)$ plot, an estimated value of N has to be used to obtain $1/G$ from Fig. 4. For not too large values of σ_B/σ_∞, the effect of N on $1/G$ is small, and therefore an estimated value of N is exact enough. Alternatively, an iterative procedure can be applied.

Bending Tests with Constant Bending Moment

General Relations

From Eq 8 we can derive a lower and upper limit of creep velocity, since

$$\frac{\dot{\sigma}}{E} + B\sigma^n \le \dot{\epsilon} \le \frac{\dot{\sigma}}{E} + \alpha C\sigma^n + B\sigma^n$$

with the strain rate limit

$$\dot{\epsilon}_L = \frac{\dot{\sigma}}{E} + B\sigma^n \qquad (18)$$

resulting from Eq 8c for $\alpha t \gg 1$

$$\dot{\epsilon}_u = \frac{\dot{\sigma}}{E} + D\sigma^n, \qquad D = \alpha C + B \qquad (19)$$

resulting from Eq 8c for $\alpha t \ll 1$.

Both special cases are of the same type, so one can restrict oneself to Eq 19.

Multiplying this equation by the distance x from the neutral axis, integrating over the height of a bending bar, and taking into account that

$$\int \frac{\dot{\sigma}}{E} x dx = 0 \qquad (20)$$

leads to

$$\dot{\epsilon}(y) = 3yD \int_0^1 \sigma^n y dy$$

with

$$y = \frac{x}{h/2}$$

Now we are using a normalized time, τ, and a normalized stress, S

$$\tau = \sigma(0)^{n-1} DEt$$

$$(21)$$

$$S = \frac{\sigma}{\sigma(0)}$$

with the elastic outer fiber stress $\sigma(0) = \sigma(t = 0, y = 1)$.

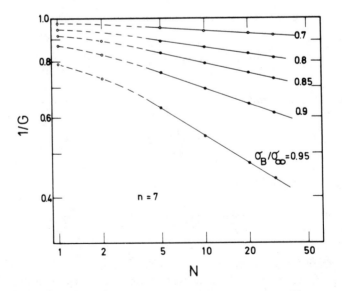

FIG. 4—*Stress rate correction factor for tests with constant displacement rate versus crack growth exponent, N.*

From Eqs 19 and 20, one obtains

$$\frac{dS}{d\tau} = -S^n + 3y \int_0^1 S^n y\, dy \qquad (22)$$

The numerical solution of Eq 22 leads to the normalized stress S as a function of normalized time, τ, and the distance, y, from the neutral axis. The outer fiber stress is plotted versus normalized time in Fig. 5. The lower curve is for the beginning of primary creep (Eq 19). The upper curve is for stationary creep (Eq 18). The real curve starts at the lower curve and shifts to the upper curve.

A calculation of this curve can be done by applying the time hardening relationship of Eq 8c.

Therefore a new "time quantity"

$$\tau^* = [C(1 - e^{-\alpha t}) + Bt] \cdot \sigma(0)^{n-1} E$$

$$\qquad (23)$$

$$= \frac{\gamma}{R} (1 - e^{-\tau R}) + (1 - \gamma)\tau$$

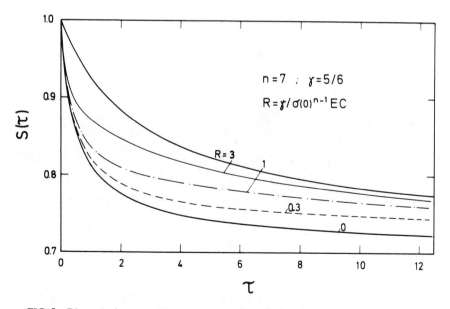

FIG. 5—*Dimensionless outer-fiber stress versus dimensionless time for tests with constant moment.*

with

$$R = \frac{\gamma}{\sigma(0)^{n-1}EC} \; ; \qquad \gamma = \frac{\alpha C}{D} \tag{24}$$

is introduced. Thus an equation similar to Eq 19 is obtained

$$\frac{d\epsilon}{d\tau^*} = \frac{1}{E}\frac{d\sigma}{d\tau^*} + \frac{\sigma^n}{\sigma_0^{n-1}E} \tag{25}$$

leading again to Eq 22, where only τ has to be replaced by τ^*.

Numerical computations are depicted in Fig. 5 for $n = 7$ and $\gamma = 5/6$. The latter is in agreement with experimental results [11,12]. Strain hardening formulation Eq 8d will give nearly the same curve, since the maximum change of bending stresses during creep is only

$$1 \leq S \leq \frac{2n+1}{3n}$$

Lifetime in Static Bending Test

From the power law of subcritical crack growth one obtains

$$dt = \frac{da}{AY^N\sigma^N a^{N/2}}$$

The outer-fiber stress, σ, is obtained from Fig. 5 and can be written in the form

$$\sigma = \sigma(0) \cdot f(t)$$

Integration leads to

$$\int_0^{t_f} f(t)^N dt = B\sigma_c^{N-2}\sigma(0)^{-N} = t_{fel} \tag{26}$$

with t_{fel} being the lifetime calculated with the linear elastic relation of Eq 4. The substitution $u = t/t_f = \tau/\tau_f$ gives

$$\tau_{fel} = \tau_f \int_0^1 f(u\tau_f)^N du \tag{27}$$

A numerical evaluation of Eq 27 is shown in Fig. 6 for different parameters, R. In Fig. 7 the effect of n is shown. The qualitative shape of curves is hardly influenced by the chosen parameters.

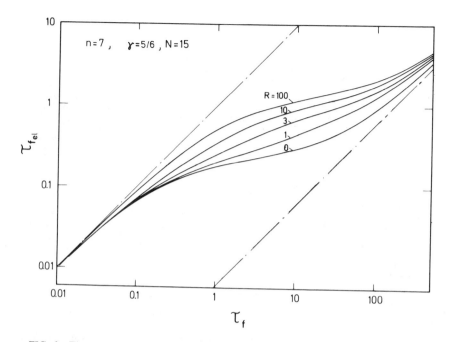

FIG. 6—*Time to failure under elastic behavior versus time to failure under creep conditions (effect of R).*

If σ_c, and therefore also t_{fel}, are given as a Weibull distribution, one obtains the creep-influenced lifetime distributions of Fig. 8.

At high loads the lifetimes, t_{fel} and t_f, are the same; at low loads lifetime scatter increases. A rough lifetime estimation, then, is given by Eq 4, where the lower asymptotic is calculated with the originally outer-fiber stress, $\sigma(0)$, and the upper asymptotic with the steady-state outer-fiber stress $\sigma = \sigma(0)(2n + 1)/3n$.

Experimental Results

Bending tests were performed with hot-pressed silicon nitride (Ceranox NH 206, Annwerk GmbH, Rödental, Germany) containing 2.5 wt% magnesium oxide (MgO) and with a density of 3.20 g/cm³. Specimens of 3.5 by 5 by 45 mm were diamond machined from plain parallel billets and annealed 4 h at 1200°C in a vacuum of 10^{-5} torr. Then dynamic and static four-point bending tests (20 mm inner, 40 mm outer span) were conducted in a furnace.

Figure 9 illustrates the conventional bending strength, $\sigma_{Be} = M_B/W$, and outer-fiber stress at failure, σ_B, as a function of stress rate for several temperatures. Each point corresponds to at least twelve measurements. $\dot{\sigma}_e$ is the

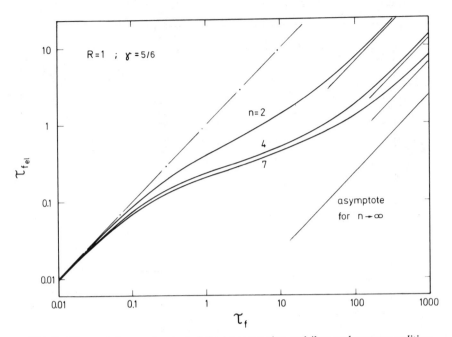

FIG. 7—*Time to failure under elastic behavior versus time to failure under creep conditions (effect of n).*

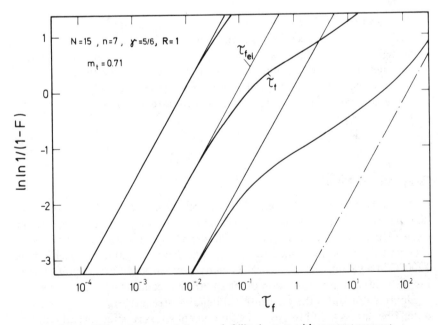

FIG. 8—*Effect of creep on failure probability for tests with constant moment.*

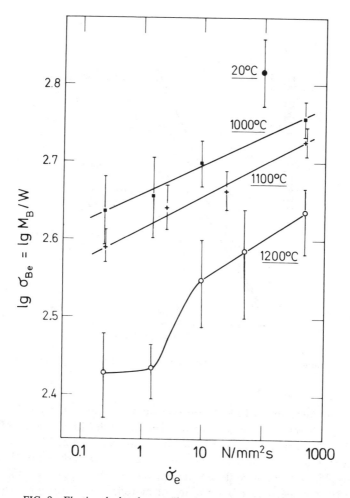

FIG. 9—*Elastic calculated outer-fiber fracture stress versus stress rate.*

stress rate in the originally linear region of bending moment versus deflection curve. At 1000°C these curves were ideal linear for all stress rates applied; at 1100°C nonlinear behavior was observed only for the lowest stress rate.

At 1200°C curves were crooked, especially at $\dot{\sigma}_e < 10 \, MN/m^2 s$. Therefore, it was necessary to make use of Eq 17 in evaluating N at 1100°C and 1200°C. In Fig. 10 the effect of the displacement rate on the shape of the bending moment–displacement curve is shown.

For the lowest loading rate at 1100°C the applied correction with $n = 7$ yielded $\sigma_B = 380 \, N/mm^2$ compared to $M_B/W = 391 \, N/mm^2$, and in $1/G = 0.77$. With this corrected value, $N = 22.6$ obtained compared to $N = 23.2$ when the uncorrected value had been used.

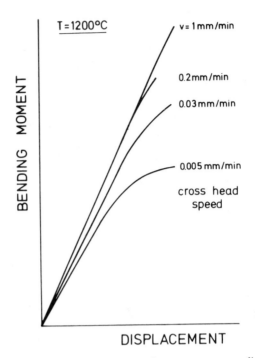

FIG. 10—*Bending moment versus displacement for tests at constant displacement rate at 1200°C.*

For the temperature of 1200°C, the effect of the correction is shown in Fig. 11. The open circles are the elastically calculated outer-fiber stresses plotted versus $\dot{\sigma}_e$. The straight line is the maximum possible outer-fiber stress σ_∞ plotted versus $\dot{\sigma}_e$. The effect of the stress rate correction is shown by the open squares. This correction could not be applied for the lowest loading rates, since corrections become a factor of ten or more in $\dot{\sigma}$, and was not necessary for the highest loading rate.

Figure 12 again shows the corrected results for the three highest loading rates and least-square fitted straight line, resulting in an N of 14.5.

Figure 13 shows Weibull plots at three temperatures and two different stress rates. The Weibull parameters m and m' are given in Table 1 for 20°C, 1000°C, and, for the highest loading rate, 1200°C. The same m as found in Table 1 is obtained for 20°C, 1000°C, and 1200°C, so one can tentatively conclude that in all of these tests with high loading rate failure is caused by the same flaw distribution. This is in contrast to observations by Wiederhorn and Tighe [7], who observed changing in flaw population during long exposure at high temperature.

On the other hand, scattering of strength decreases in case of dominating creep behavior. For $\dot{\sigma}_e = 1\ MN/m^2 s$ $m = 22.5$ was experimentally determined at 1200°C, in contrast to the short time values of Table 1.

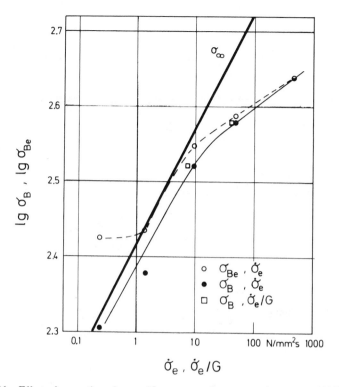

FIG. 11—*Effect of correction of outer-fiber stress and stress rate for tests at 1200°C. Dark line: limit-bending moment under creep conditions versus stress rate, $\dot{\sigma}_e$.*

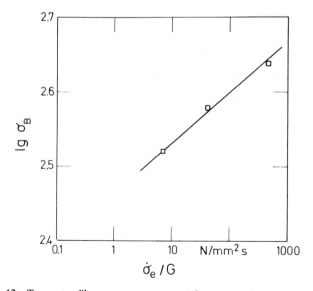

FIG. 12—*True outer-fiber stress versus corrected stress rate for tests at 1200°C.*

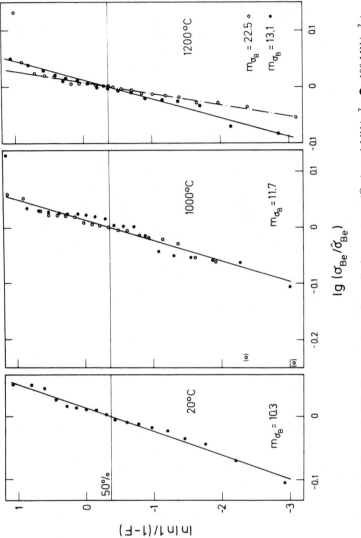

FIG. 13—*Failure probability for tests with constant displacement rate* ($\bigcirc$, $\dot{\sigma}_e = 1 \; MN/m^2 s$; $\bullet$, $100 \; MN/m^2 s$).

TABLE 1—*Results from dynamic bending tests.*

	Temperature			
	20°C	1000°C	1100°C	1200°C
N	...	25.8 ± 2.1	22.6 ± 2.5	14.5 ± 1.3
log $\hat{\sigma}_B$ (1 N/mm^2s)	...	2.657 ± 0.0045	2.611 ± 0.007	2.463 ± 0.010
m' (measured)	10.3	11.7	...	13.1
$m = m' \dfrac{N-2}{N+1}$	10.3	10.4	...	10.5

In Fig. 14 the results are shown for the tests with constant moment. The failure probability is plotted versus the time to failure in a Weibull diagram. The predictions using linear elastic relations are also shown. At 1000 and 1100°C there is a good agreement between linear elastic predictions and measurements.

At 1200°C, predictions were made with the linear elastic relation in Eq 4, as well as with the creep-corrected relation in Eq 27. Both results are given in Figs. 14c and 15. Linear elastic evaluation gives a good agreement for short lifetimes, that is, high stresses. Consideration of the creep behavior gives an at least qualitative agreement. It seems that there is a shift between prediction and measurement with a constant factor of about two to three in the noncon-

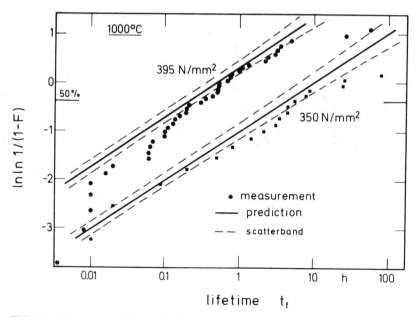

FIG. 14a—*Measured and linear elastic predicted failure probability for tests with constant moment, at 1000°C.*

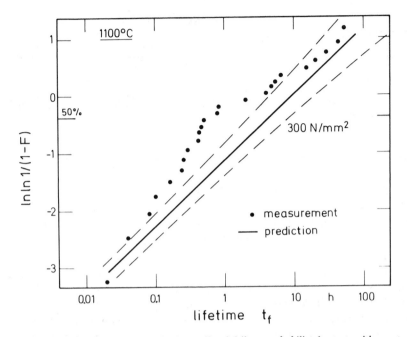

FIG. 14b—*Measured and linear elastic predicted failure probability for tests with constant moment, at 1100°C.*

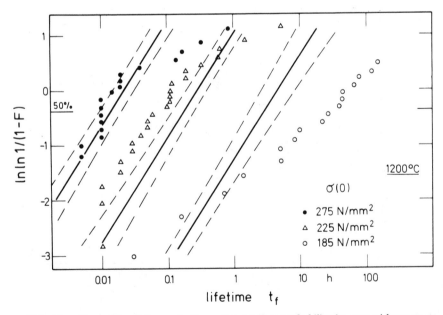

FIG. 14c—*Measured and linear elastic predicted failure probability for tests with constant moment, at 1200°C.*

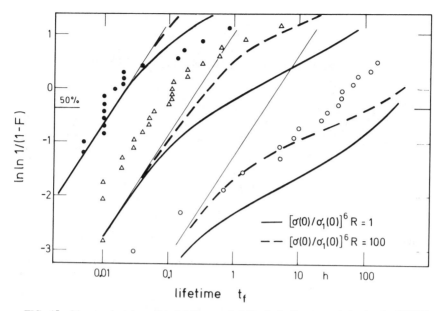

FIG. 15—*Measured and predicted failure probability including creep behavior for 1200°C; $\sigma_l(0) = 225 \, N/mm^2$.*

servative direction. This can be caused by small differences in temperatures between dynamic and static bending tests.

Conclusions

Specimens of hot-pressed silicon nitride were tested in four-point bending at constant displacement rate and with constant load at elevated temperatures. Subcritical crack extension resulted in delayed fracture for the constant load tests and an increase in fracture strength for the tests performed at constant displacement rate. A quantitative explanation of the effects including the scatter of the rupture time was possible using linear elastic fracture mechanics, however, taking into account the changing stress distribution in the specimen due to creep. The following results were obtained:

1. For tests performed at 1000 and 1100°C the failure probability–failure time relationship for the constant load tests could be predicted from tests with constant displacement rate.

2. The stress exponent of the creep relation of Pao and Marin can be obtained from four-point bend tests carried out at constant displacement rate.

3. To determine the exponent N in the relationship between crack velocity and stress intensity factor under creep conditions, relations were obtained for correcting the outer-fiber stress and the stress rate.

4. The same Weibull parameter, m, was obtained at 20, 1000, and 1100°C

and also at 1200°C for high deflection rates. At 1200°C and a low deflection rate, a higher parameter m was observed.

5. The relationship between failure probability and time to failure for tests performed at constant load could be qualitatively predicted by taking into account the changing outer-fiber stress due to creep.

Acknowledgments

The experimental work was conducted while one of the authors (Fett) was affiliated with the German Aerospace Research Establishment (DFVLR). We gratefully acknowledge the assistance of J. Eschweiler and K. Spiegel in performing the experiments. The work was supported by the Deutsche Forschungsgemeinschaft.

References

[1] Evans, A. G. and Wiederhorn, S. M., "Proof Testing of Ceramic Materials—an Analytical Basis for Failure Prediction," *International Journal of Fracture*, Vol. 10, 1974, pp. 379–392.

[2] Ritter, J. E., Jr., "Engineering Design and Fatigue Failure of Brittle Materials," *Fracture Mechanics of Ceramics*, Vol. 4, Plenum, New York, 1978, pp. 667–686.

[3] Jakus, L., Coyne, D. C., and Ritter, J. E., Jr., "Analysis of Fatigue Data for Lifetime Predictions for Ceramic Materials," *Journal of Materials Science*, Vol. 13, 1978, pp. 2071–2080.

[4] Richter, H. and Soltész, V., "Proof-Test an 3-Punkt-Biegeproben aus Al_2O_3-Keramik," 12, Sitzung AK "Bruchvorgänge," *Deutscher Verband für Materialprüfung*, Berlin, 1980, pp. 105–110.

[5] Fett, T., "Lebensdauervorhersage an keramischen Werkstoffen mit den Methoden der Bruchmechanik bei elastischem und viskoelastischem Materialverhalten," Thesis, Universität Karlsruhe, Karlsruhe, Germany, 1983.

[6] Trantina, G. G., "Strength and Life Prediction for Hot-Pressed Silicon Nitride," *Journal of the American Ceramic Society*, Vol. 62, 1979, pp. 377–380.

[7] Wiederhorn, S. M. and Tighe, N. J., "Proof-Testing of Hot-Pressed Silicon Nitride," *Journal of Materials Science*, Vol. 13, 1978, pp. 1781–1793.

[8] Lange, F. F., "Interrelations Between Creep and Slow Crack Growth for Tensile Loading Conditions," *International Journal of Fracture*, Vol. 12, 1976, pp. 739–744.

[9] Charles, R. J. "Dynamic Fatigue of Glass," *Journal of Applied Physics*, Vol. 29, 1958, pp. 1657–1661.

[10] Pao, Y. H. and Marin, J., "An Analytical Theory of the Creep Deformation of Materials," *Journal of Applied Mechanics*, Vol. 20, 1953, pp. 245–252.

[11] Kossowsky, R., Miller, D. G., and Diaz, E. S., "Tensile and Creep Strength of Hot-Pressed Si_3N_4," *Journal of Materials Science*, Vol. 10, 1975, pp. 983–997.

[12] Salah ud Din and Nicholson, P. S., "Creep of Hot-Pressed Silicon Nitride," *Journal of Materials Science*, Vol. 10, 1975, pp. 1375–1380.

[13] Lange, F. F., "High Temperature Deformation and Fracture Phenomena of Polyphase Si_3N_4 Materials," in *Progress in Nitrogen Ceramics*, F. L. Riley, Ed., M. Nijhoff Publishers, Boston, 1983, pp. 467–490.

[14] Arons, R. M. and Tien, J. K., "Creep and Strain Recovery in Hot-Pressed Silicon Nitride," *Journal of Materials Science*, Vol. 15, 1980, pp. 2046–2058.

[15] Evans, A. G. and Wiederhorn, S. M. "Crack Propagation and Failure Prediction in Silicon Nitride at Elevated Temperatures," *Journal of Materials Science*, Vol. 9, 1974, pp. 270 ff.

[16] Shih, T. T. and Opoku, J., "Application of Fracture Mechanics to Ceramic Materials—A State of the Art Review," Scientific Paper 78-9D3-CTRDP-P1, Westinghouse R&D Center, Pittsburgh, 1978.

George D. Quinn[1]

Static Fatigue in High-Performance Ceramics

REFERENCE: Quinn, G. D., "Static Fatigue in High-Performance Ceramics," *Methods for Assessing the Structural Reliability of Brittle Materials, ASTM STP 844,* S. W. Freiman and C. M. Hudson, Eds., American Society for Testing and Materials, Philadelphia, 1984, pp. 177–193.

ABSTRACT: Static fatigue failure in ceramics is often assumed to be due to one mechanism: slow crack growth from preexisting flaws. Multiple mechanisms of failure may be operative, however. Stress rupture data for two materials, hot-pressed silicon nitride and sintered silicon carbide, will be presented and interpreted in terms of multiple mechanisms.

KEY WORDS: static fatigue, ceramics, crack growth, strength, materials fracture, delayed failure, stress rupture, creep fracture, structural reliability, brittle materials

Static fatigue phenomena in ceramics can be detected by many methods, including dynamic fatigue, double torsion, and double-cantilever beam testing. These methods either directly or indirectly measure growth rates of preexisting flaws or large cracks. Fracture mechanics analysis can be used to interpret the results and apply them to static fatigue. Unfortunately, these methods are not foolproof since they measure only fast-moving cracks in specimens loaded nearly to fracture. Uncertainties in the crack growth function, and the need to extrapolate can render data from these techniques unreliable for lifetime prediction [1]. These analyses are pertinent only where failure occurs by the growth of preexisting flaws or artificial cracks. Static fatigue may occur from other mechanisms.

Stress rupture testing can detect static fatigue phenomena over a much wider range of conditions and does not presuppose any mechanism of failure. Stress rupture entails the application of a constant load to a specimen. Failure occurs by whatever mechanism is dominant for a given set of conditions. Analysis is simple and has a minimum number of assumptions. The flexural mode

[1]Ceramic engineer, U.S. Army Materials and Mechanics Research Center, Watertown, Mass. 02172.

of loading is commonly used because of its experimental facility. Tension stress rupture data on high-performance ceramics are extremely rare, but recent work has appeared which lends insight into the value of flexure testing.

The paper will focus on the two high-performance ceramics, hot-pressed silicon nitride (HPSN) (NC-132 grade, Norton Co., Worcester, Mass.) and sintered alpha silicon carbide (SASC) (Carborundum Co., Niagra Falls, N.Y.), for which the greatest body of data exists. The presence of multiple mechanisms of static fatigue failure will be discussed. Stress rupture testing has discerned these multiple mechanisms, whereas the other methods have not.

The terminology used in this paper needs to be defined, since there are subtle variations in meaning between the expressions. Static fatigue is the most general expression applicable to failure of a specimen at a finite time after a load is applied. Stress corrosion, which is a mechanism by which static fatigue can occur, refers to the growth of cracks by chemical attack in stressed specimens. Stress corrosion may involve nucleation of cracks as well. Creep fracture, a different mechanism by which static fatigue can occur, pertains to the formation of extensive creep microcrack networks which coalesce, eventually leading to fracture. Another mechanism is slow crack growth (SCG), which applies to the extension of preexisting flaws. Although slow crack growth may involve creep deformation and microcracking local to the crack tip, it does not involve bulk creep and microcracking typical of creep fracture. SCG is not necessarily environmentally dependent, which distinguishes it from stress corrosion.

Most life prediction analyses to date have interpreted stress rupture results in terms of SCG from preexisting flaws. A power function relating crack velocity and stress intensity

$$V = AK_I^n \qquad (1)$$

where V is crack velocity, A and n are constants, and K_I is stress intensity, is often used to develop equations for time to failure. This equation is a convenient form but is empirical and as yet has no theoretical foundation. Indeed, it may be irrelevant in those instances in which static fatigue failure is due to creep fracture.

Hot-Pressed Silicon Nitride

NC-132 HPSN is gaining wide acceptance as a model structural ceramic that is vulnerable to static fatigue. More stress rupture data exist for this material than for any other high-performance ceramic. Figure 1, which is from Ref 2, illustrates the unusually consistent behavior of this material at 1200°C. There is excellent agreement among the flexural results despite different source billets, specimen sizes, specimen preparations, and test fixtures. Figure 2 [3] also shows the consistency of this material at temperatures over 1000°C. These re-

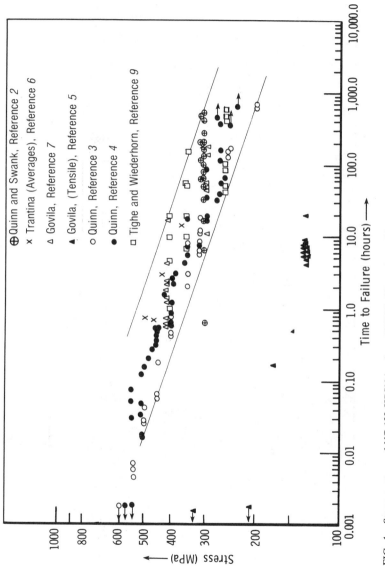

FIG. 1—*Stress rupture of NC-132 HPSN in air at 1200°C. The points with arrows are either immediate failures or survivors.*

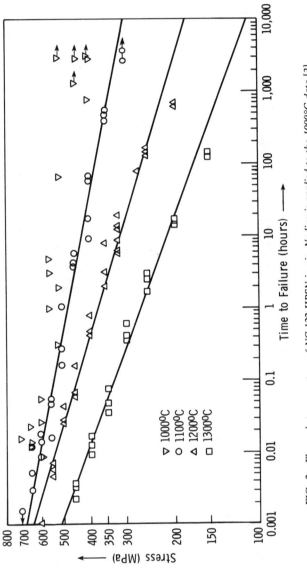

FIG. 2—*Flexural stress rupture of NC-132 HPSN in air. No line is applied to the 1000°C data* [3].

TABLE 1—*Fast fracture normalization factors.*

References	σ_0,MPa	Comments
Quinn [4]	714 825	Room temperature, four-point flexure, 3.05-cm span; two lots of specimens, the appropriate σ_0 was used for each datum
Quinn [3]	909	Room temperature, four-point flexure, 3.05-cm span
Govila [7]	736	Room temperature, four-point flexure, 1.91-cm span
Govila [5]	250	1204°C tension; extrapolation of stress rupture data back to 0.001 h since no fast fracture data available
Trantina [6]	873	Room temperature, three-point flexure 2.3-cm span

sults will now be examined, particularly with respect to whether failure is due only to SCG from preexisting flaws.

Trials longer than 1000 h at 1200°C or higher are often not practical because of excessive creep deformations, which can cause specimens to bottom out on fixturing. Care must be taken in the interpretation of the flexural data, despite the excellent fit of the lines to the data. Creep deformation can alter the stress distributions in the specimens. All the flexural stresses were calculated from the elastic solution. The actual peak tensile stress may be considerably less; the longer the specimen is under load, the greater the reduction will be. Short-time-to-failure specimens may not be affected very much. Thus, the elastic stresses listed in Figs. 1 and 2 are upper limits. If the corrections to stress were made, the lines in Figs. 1 and 2 would have steeper downward slopes. The scatter in failure time for identically loaded samples at 1200 or 1300°C is very low compared with most high-performance ceramics. Reference *4* contains an analysis which attempts to account for the scatter.[2] A 29% difference in time was possible due to a (worst case) 3% error in stress; a 22% variation would occur due to a possible 5°C temperature error, but the worst variability—a factor of over 300—should have resulted from initial flaw size variations. The latter factor was determined by scatter of room-temperature fast fracture strengths. That the true scatter was on the order of 10 to 15, implies that initial flaws were not particularly important in influencing time to failure.

A closer examination of Fig. 1 suggests there is a slight difference (by a factor of two to five) in lifetime between the data sets. The curves might be expected to extrapolate back (at short times) to the fast fracture strength. A graph showing strength loss with time, in relation to the fast fracture strength, could improve the agreement between the data sets. Figure 3 shows the data with the applied stresses normalized by the appropriate fast fracture strength, which should have factored out differences due to specimen size, mode of loading, billet sources, and preparation procedures. Table 1 [3–7] gives details on the normalization strengths. The normalization was not successful, and one of

[2]The first printing of Ref *4* contained several typographical errors, but the calculations were correct.

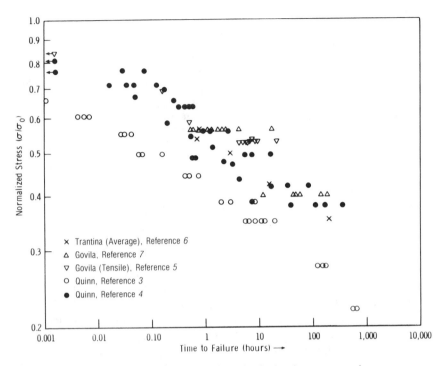

FIG. 3—*The data of Fig. 1 normalized by the fast fracture strength, σ_0.*

the data sets falls significantly off the trend in Fig. 3. It is not clear whether the data sets extrapolate back to 1.0 normalized stress at short failure times. A key assumption of the normalization approach which may be violated is that crack growth occurs from the same defects that cause fast fracture failures. Furthermore, the room temperature strength does not incorporate variable chemistry effects that may be significant at elevated temperatures. Finally, if failure is due to creep fracture, then the initial flaw distribution will be of no consequence.

Govila has performed tensile stress rupture on NC-132 HPSN [5, 7]. This work is especially valuable for comparison to the flexural data. The 1200°C results are included in Fig. 1. Although the curve is not complete, it is evident that the trend is similar to that of the flexure data. That the tension specimens failed for much smaller applied stresses may not be surprising in view of statistical theories of strength, which incorporate specimen size effects. Depending upon whether an effective surface or volume analysis is used, and whether Govila's elevated or room-temperature fast fracture strength data are used, Weibull theory predicts a strength ratio (flexure to tension) between 1.1 and 1.5. The actual ratio is about 3. Several alternative explanations can account for this discrepancy. First, the tension specimens failed because of crack growth from *volume*-distributed sources. In the flexural specimens, growth may have been from *surface* origins. Second, creep relaxations, known to be significant at 1200°C, could reduce the effective stresses in the flexure specimens. Third, it is possible that differences in the magnitude of the applied stresses could change the mechanism of static fatigue. The higher applied stresses in flexure could lead to SCG from preexisting flaws, while the lower stresses in tension could accentuate creep fracture. Finally, if the mechanism of failure is indeed creep fracture, the initial flaws will be of no consequence, whereupon the Weibull approach is irrelevant.

It was similarly noted in Ref 4, that although all flexural fast fracture specimens broke from surface defects, at least several stress rupture specimens failed because of volume-originated crack growth. In fact, the short-time-to-failure specimens, and those tested at 1000°C, had little creep deformation and fracture surfaces with small SCG zones. The higher the temperature and the longer the time to failure, the more likely that failure was related to creep fracture. Thus, the curves of Figs. 1 and 2 may reflect two overlapping mechanisms of static fatigue failure. Unfortunately, it is difficult to determine from fracture surfaces whether failure was due to SCG from an initial flaw or to creep fracture.

Perhaps the strongest evidence that under some conditions initial flaws are of no consequence has been the findings of studies with artificially flawed specimens. A Knoop (or occasionally Vickers) indentation is used to implant not only a hardness indent but also a subsurface crack. The intent is to create a flaw of dimensions large enough to guarantee that it is the most severe in the specimen. By varying the indentation loads, it is possible to control the initial flaw size. Providing it is large enough, fast fracture will always originate at the artificial flaw.

The most comprehensive stress rupture study with artificially flawed specimens is reported in Ref 3. Test conditions in that study were carefully chosen to accentuate failure from SCG, since it was desired to develop life prediction parameters appropriate to that mechanism. Most specimens had 1.6 kg Knoop flaws (50×10^{-6} m deep cracks) and were loaded at 266 MPa. Subsequent to that study, an additional 17 specimens were tested in the same manner, but over a wider range of stresses.[3] Complete details of the experimental procedure are in Ref 3. The results are shown in Fig. 4, along with the earlier data. Each labeled point in Fig. 4 corresponds to a stress-temperature condition at which stress rupture experiments were performed. The points are labeled by a ratio of the number of specimens that *did* fail from the artificial flaw to the *total* number tested. The number in parenthesis is the approximate median time to failure. For example, at 266 MPa and 1250°C, of 13 specimens tested, 10 broke at the flaw, and the median time to failure was 5 h. The artificial flaws limited the fast fracture strength to less than 400 MPa for most tem-

---

[3]All specimens were from Billet A, referred to in Ref 3.

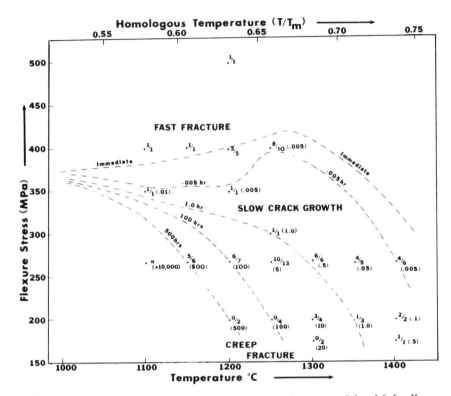

FIG. 4—*Fracture map for flexural stress rupture specimens containing 1.6 kg Knoop artificial flaws. The symbols are explained in the text.*

peratures. This confirms the dominance of the artificial flaws over the natural flaws. Naturally flawed specimens have strengths in the 500 to 1000 MPa range [3]. From 200 to 300 MPa, most specimens failed because of SCG from the artificial flaws. Those that did not may either have had a larger natural flaw or have failed because of creep fracture. Loci of constant failure time are shown as dotted lines in Fig. 4. Point a, which is at 266 MPa at 1100°C, refers to two trials underway at U.S. Army Materials and Mechanics Research Center (AMMRC) which have over 10,000 h accumulated as of November 1982. At 200 MPa or lower, creep fracture was dominant, since creep deformation became significant and a high fraction of specimens did not fail at the artificial flaw. Thus, fractographic examination *was* able to delineate the mechanisms of failure for the artificially flawed specimens. (It is not possible to do so for naturally flawed specimens.) Although there are too few specimens to establish definitive boundaries, it is clear that there are different behavior regimens. Caution must be exercised in generalizing these results since they were generated for artificially flawed flexural specimens. A map for naturally flawed specimens or tension specimens could be very different.

Creep mechanism maps are well known, but fracture maps are relatively recent. Gandhi and Ashby [8] have presented a preliminary diagram for an earlier vintage HPSN (HS-130 grade, Norton Co., Worcester, Mass., or Joseph Lucas Co., Solihall West Midland, England). Their terminology for the slow crack growth regimen is "brittle intergranular fracture 3." They used normalized temperature and stress axes. The temperature axis was a homologous temperature, with the T_m for HPSN set at 1900°C. Figure 4 of this paper includes such a normalized axis. The normalized stress axis used by Gandhi and Ashby is not shown, since it ignores differences such as flexure versus tension testing and artificial versus natural flaw. Preoxidation can affect whether specimens break from the artificial flaws [3,9,10] and can have a profound effect upon lifetime [2]. Nevertheless, it is evident that with further experimentation and analysis Fig. 4 could evolve into a generalized fracture mechanism map.

Life prediction equations for SCG from preexisting flaws and creep fracture will surely differ. Initial flaw size is a variable for the former but is irrelevant to the latter. Creep rate and total strain at failure will be important parameters that can be used to predict lifetime under creep fracture conditions. At many stresses and temperatures, it is probable that both mechanisms will be operative simultaneously, and it will be necessary to determine which is dominant.

Sintered Alpha Silicon Carbide

Sintered Alpha Silicon Carbide (SASC) is also susceptible to static fatigue, but usually only for loads 65% or greater than its fast fracture strength. Typical results in Figs. 5 and 6 show a considerable scatter in time to failure for identically loaded specimens. SASC is much more resistant to creep than HPSN [16] and therefore is less susceptible to creep fracture. Stress rupture experiments have confirmed this, and no instances of creep fracture have been reported at temperatures below 1500°C [6,11-15,17,18]. On the other hand, there is evidence that more than one mechanism of static fatigue is operative, depending upon the stress and temperature regimen. Intergranular SCG has been a cause of static fatigue failure at temperatures greater than 1200°C, but at lower temperatures, an alternate mechanism involving surface-connected porosity can cause failure. A clear definition of the stress-temperature regimens is difficult because SASC has shown some variability in behavior through the years [11] and also because it is difficult to identify the low-temperature mechanism.

The SCG is readily discernible through fractography, since it is intergranular whereas fast fracture is transgranular (Fig. 7). Srinivasan and his colleagues [15,17] determined that SCG caused static fatigue failures in specimens (nearly loaded to fast fracture) that contained artificial Knoop flaws. This occurred at 1400 and 1500°C, and although there was some crack growth

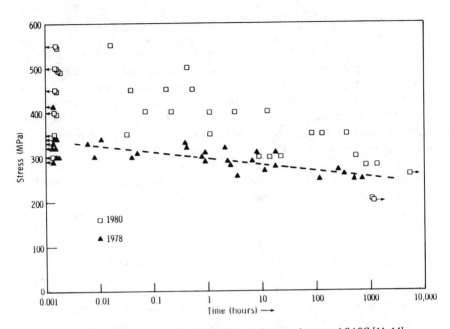

FIG. 5—*Flexural stress rupture at 1200°C in air for two vintages of SASC* [11–14].

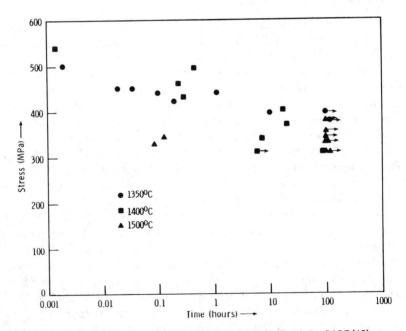

FIG. 6—*Flexural stress rupture at 1350°C and higher in air for SASC* [15].

FIG. 7—*Two views of a fracture surface of an SASC flexural specimen that failed at 1400°C. The entire SCG zone is shown in the top photograph. A closeup of the intergranular-transgranular boundary between SCG and fast fracture is shown in the bottom photograph* [4].

at 1200°C, it was negligible and did not lead to failure within 100 h. Quinn [4], and more recently Govila [18], observed SCG in naturally flawed flexure specimens at 1300 and 1400°C. Govila's tension stress rupture experiments also identified SCG as the cause of failure at 1300°C in air [18]. The SCG zones in all these instances were substantially smaller (in relation to the specimen size) than those that occur in HPSN.

The low-temperature mechanism is usually (but not always) observed below 1300°C. It is much more difficult to identify since fracture surfaces often appear identical to room temperature specimens. Figure 8 shows the origins for an SASC room temperature fast fracture specimen (408 MPa) and a stress rupture specimen that failed at 2.1 h at 1200°C in air while sustaining a 320-MPa stress. Often the origin of time-dependent failure is not a discrete pore, but is apparently a zone of less dense, or not fully sintered, material, as is shown in Fig. 9. SASC is not fully dense material and ordinarily contains grain boundary pores throughout the bulk. It is difficult to determine if some of the features are intergranular cracking. The low-temperature mechanism may be a stress corrosion phenomenon. Quinn has reported that time-dependent failures (in three vintages of SASC) at 1200°C originated almost exclusively in surface-connected pores or porous zones [4,12-14]. This was unlike fast fracture or SCG time-dependent fractures, which had a variety of origins, including surface and internal porosity, machining damage, and large grains. The static fatigue failures occurred at temperatures as low as 1000°C. Govila's tension stress rupture specimens also had surface-connected origins in every case at 1200 and 1300°C, unlike his earlier work on HPSN [18]. Srinivasan performed flexure experiments in air and argon and had no static fatigue failure in the inert atmosphere, whereas numerous failures did occur in air [15]. Srinivasan interpreted the low-temperature mechanism to be an environmentally assisted, localized creep cavitation process. Thus, although there are conflicting interpretations regarding the identity of the low-temperature mechanism, it is certain that it is environmentally assisted and is confined to an extremely small zone around an origin, which is nearly always a surface-connected pore or porous zone.

These difficulties in identifying the low-temperature mechanism, and distinguishing it from limited SCG, may be attributable in part to variability in SASC through the years. Quinn has presented three different static fatigue curves at 1200°C, depending upon whether the material was a 1977, 1978, or 1980 vintage [4,12-14]. Srinivasan, in his recent work [15], reported that naturally flawed specimens failed from surface-connected flaws that did *not* exhibit SCG,[4] unlike his earlier studies [17] with artificially flawed specimens which *did* fail from SCG. Either there is a vintage effect, or the artificially flawed specimens which had relatively low stresses on them were susceptible to

---

[4]This study was the only instance in which the alternate (to SCG) mechanism occurred at a temperature over 1300°C.

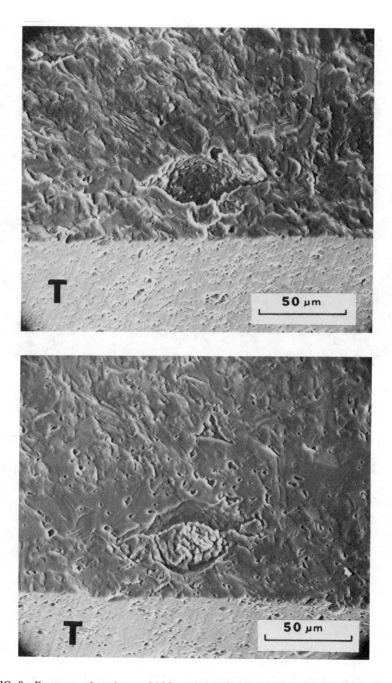

FIG. 8—*Fracture surfaces for two SASC specimens showing (top) a pore that was of a room-temperature fast-fracture origin and (bottom) a pore which was of a 1200°C stress rupture origin. The flexural specimens were tilted back to show the tensile surface, "T" [4].*

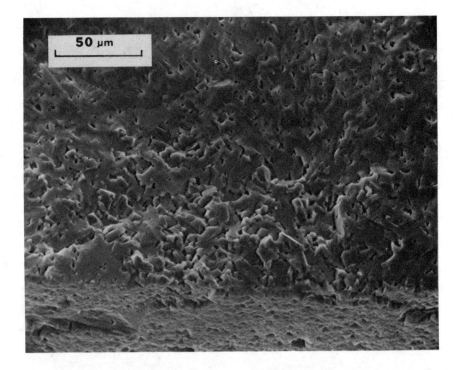

FIG. 9—*Failure origin in an SASC flexural specimen which failed at 1461 h at 1200°C in air* [14].

SCG while naturally flawed specimen were prone to fail from the alternate mechanism.

Govila's tension stress rupture experiments [18] have again been a valuable benefit in the interpretation of flexural results. Static fatigue trends and phenomena observed with flexure have been substantiated by the tensile tests. Figure 10 shows the results for the two methods at 1300°C. The trends are similar and the loci of failure are shifted by a factor of about 1.4 in stress. A Weibull surface correlation predicts a shift of only 1.1 if the room-temperature fast fracture modulus (10.9) is used.[5] The surface analysis is appropriate, since nearly all time-dependent failures were surface or near-surface originated, for both flexure and tension. Although the factors are in better agreement than the HPSN values, it is still of interest to carefully examine the difference. Unlike HPSN, creep is not a problem. The room temperature modulus reflects a variety of flaws, including volume and surface porosity, but the high-temperature stress rupture failures involved only surface-connected pores. Thus the room temperature Weibull modulus number may be inappropriate. More serious, perhaps, the flexural fixture used in the study may have had experimental

[5]A volume analysis (which ignores the fractographic evidence) gives a ratio of 1.3.

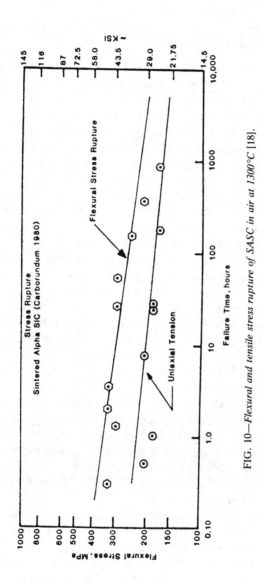

FIG. 10—Flexural and tensile stress rupture of SASC in air at 1300°C [18].

errors on the order of tens of percent.[6] This uncertainty points out the desirability of minimizing experimental errors in flexure testing and developing standard test methods. A paper presented in this volume will specifically deal with these issues [19].

Summary and Conclusions

The existence of a multitude of mechanisms of static fatigue failure has been demonstrated for two high-performance ceramics. HPSN is susceptible to SCG from preexisting flaws and creep fracture. High temperature and low stress promote the latter. Further studies should be focused upon refinements of the fracture map (Fig. 4) and developing different life prediction equations for the appropriate mechanisms. More attention should be given to creep rate and strain at failure for predictions of creep fracture lifetime.

SCG and creep fractures are related to the creep resistance of the material. The difference is a matter of scale. In the former, creep deformations are confined to the immediate vicinity of a crack tip and lead to crack extension. Alternatively, if there is significant deformation throughout the material, creep fracture will result. Although there are known billet-to-billet variations in its static fatigue resistance, HPSN nonetheless is a relatively consistent material.

SASC can fail in static fatigue because of SCG from preexisting flaws, but it is also susceptible to an undefined alternate mechanism. The SCG failures are usually observed above 1300°C and the alternate mechanism below that temperature. The latter is environmentally dependent and is confined to an extremely small zone, usually at a surface pore or porous zone. The SCG zones are much larger, but nevertheless are smaller than those observed in HPSN. Unlike HPSN, static fatigue in SASC is confined to stresses greater than 65% of the fast fracture strength. SASC does not have the same consistency in static fatigue as HPSN. It is premature to develop a fracture map for SASC.

Multiple causes of static fatigue failure may be present in other high-performance ceramics as well. SCG and phase instability can lead to static fatigue in yttria-doped HPSN [11]. Reaction sintered silicon nitride may be susceptible to creep fracture over 1400°C, an environmentally assisted, surface-pore-related phenomenon between 1000 and 1400°C, and also to stress corrosion (involving water) at room temperature [11].

Most models of static fatigue assume the material is homogeneous. This is not necessarily so. Surface-connected pores appear to be especially vulnerable locations for static fatigue failure in high-performance ceramics. Furthermore, it is well known that heat treatments and oxidation can alter the structure and chemical distributions within a material. For example, oxidation can increase the amount and alter the composition of grain boundary phases, particularly at the surface. Zones of differing vulnerability to static fatigue may exist within a component or specimen, and they may change with time.

---

[6]Major errors arise from load-bearing friction and specimen twisting.

Testing methods have an important influence upon static fatigue detection and characterization. Stress rupture methods are clearly superior to fracture mechanics tests. Tension testing, while difficult and expensive, is extremely valuable in substantiating the flexural results. It is not likely that enough tension data will ever be available to serve as a design base, so flexure may have to suffice. Improvements in flexure testing are essential before meaningful design data can be generated.

References

[1] Wiederhorn, S. and Ritter, J., Jr., "Application of Fracture Mechanics Concepts to Structural Ceramics," *Fracture Mechanics Applied to Brittle Materials, ASTM STP 678*, American Society for Testing and Materials, Philadelphia, 1979, pp. 202-214.

[2] Quinn, G. and Swank, L., *Communications of the American Ceramic Society*, January 1983, pp. C31-C32.

[3] Quinn, G. and Quinn, J. "Slow Crack Growth in Hot-Pressed Silicon Nitride," in *Fracture Mechanics of Ceramics*, Vol. 6, R. C. Bradt, A. G. Evans, D. P. H. Hasselman, and F. F. Lange, Eds., Plenum, New York, 1982, pp. 603-636.

[4] Quinn, G., "Characterization of Turbine Ceramics After Long Term Environmental Exposure," TR 80-15, NTIS ADA 117463, U.S. Army Materials and Mechanics Research Center, Watertown, Mass., April 1980.

[5] Govila, R., *Journal of the American Ceramic Society*, Vol. 65, No. 1, January 1982, pp. 15-21.

[6] Trantina, G., *Journal of the American Ceramic Society*, Vol. 62, Nos. 7-8, 1979, pp. 377-380.

[7] Govila, R., "Ceramic Life Prediction Parameters," TR 80-18, NTIS ADA 090272, U.S. Army Materials and Mechanics Research Center, Watertown, Mass., May 1980.

[8] Gandhi, C. and Ashby, M. F., *Acta Metallurgica*, Vol. 27, 1979, pp. 1565-1602.

[9] Tighe, N. and Wiederhorn, S., "Effects of Oxidation on the Reliability of Silicon Nitride," in *Fracture Mechanics of Ceramics*, Vol. 6, R. C. Bradt, A. G. Evans, D. P. H. Hasselman, and F. F. Lange, Eds., Plenum, New York, 1982, pp. 403-424.

[10] Das, G., Mendiratta, M., and Cornish, G., *Journal of Materials Science*, Vol. 17, 1982, pp. 2486-2494.

[11] Quinn, G., "Review of Static Fatigue in Silicon Nitride and Silicon Carbide," *Ceramic Proceedings*, Vol. 3, Nos. 1-2, 1982, pp. 77-98.

[12] Quinn, G. and Katz, R., "Time Dependence of the High Temperature Strength of Sintered Alpha Silicon Carbide," TN 79-5, U.S. Army Materials and Mechanics Research Center, Watertown, Mass., June 1979.

[13] Quinn, G. and Katz, R., "Time-Dependent High Temperature Strength of Sintered α-SiC," *Journal of the American Ceramic Society*, Vol. 63, No. 1-2, 1980, pp. 117-119.

[14] Quinn, G., "Stress Rupture of Sintered Alpha Silicon Carbide," TN 81-4, U.S. Army Materials and Mechanics Research Center, Watertown, Mass., December 1981.

[15] Srinivasan, M., "Elevated Temperature Stress Rupture Response of Sintered Alpha Silicon Carbide," (abstract), *American Ceramic Society Bulletin*, Vol. 58, No. 3, 1979, p. 347; manuscript available from the Carborundum Co., Niagara Falls, N.Y.

[16] Larsen, D., "Property Screening and Evaluation of Ceramic Turbine Engine Materials," Technical Report, Air Force Materials Laboratory TR 79-4188, October 1979.

[17] Coppola, J., Srinivasan, M., Faber, K., and Smoak, R., "High Temperature Properties of Sintered Alpha Silicon Carbide," presented at the International Symposium on Factors in Densification and Sintering of Oxide and Non-Oxide Ceramics, October 1978, Hakone, Japan; manuscript available from the Carborundum Co., Niagara Falls, N.Y.

[18] Govila, R., "High Temperature Strength Characterization of Sintered Alpha Silicon Carbide" TR 82-51, NTIS ADA 121437, U.S. Army Materials and Mechanics Research Center, Watertown, Mass., October 1982.

[19] Baratta, F., "Requirements for Flexure Testing of Brittle Materials," in this publication, pp. 194-222.

Francis I. Baratta[1]

Requirements for Flexure Testing of Brittle Materials

REFERENCE: Baratta, F. I., "**Requirements for Flexure Testing of Brittle Materials,**" *Methods for Assessing the Structural Reliability of Brittle Materials, ASTM STP 844,* S. W. Freiman and C. M. Hudson, Eds., American Society for Testing and Materials, Philadelphia, 1984, pp. 194–222.

ABSTRACT: Requirements for accurate bend testing at room temperature of four-point and three-point beams of rectangular cross section are outlined. The so-called simple beam theory assumptions are examined to yield beam geometry ratios that will result in minimum error when utilizing elasticity theory. These errors are referred to as internal error sources, whereas those errors that result from wedging stress, contact stress, load mislocation, beam twisting, friction at beam contact points, contact point tangency shift, and neglect of corner radii or chamfer geometry in stress determination are termed external error sources. Also included are estimates of errors in the determination of the Weibull parameters as a function of sample size. Error tables and formulas, based on the previously mentioned factors, are presented.

Such analyses and results provide guidance for the accurate determination of flexure strength of brittle materials within the linear elastic regimen.

KEY WORDS: test methods, flexure strength, mechanical properties, brittle materials, mechanics, structural reliability

Nomenclature

E Young's modulus of the test material

E_c Young's modulus in compression of the test material

E_T Young's modulus in tension of the test material

I Moment of inertia for a rectangular beam ($I = bd^3/12$)

L Outer span length for a four-point and a three-point loaded beam

L_T Total length of a beam

---

[1]Mechanical engineer, Army Materials and Mechanics Research Center, Watertown, Mass. 02172.

M Weibull slope parameter

M_b General moment applied to beam

P General applied force

P_1, P_2, P_3, P_4 Forces applied to a beam (see Fig. 1)

a Half the distance between the inner span and outer span for a four-point loaded beam, that is, $(L - \ell)/2$ or $a = L/2$ for a three-point loaded beam (note, $a_1 = a_2 = a$)

a_1 and a_2 A beam dimension

b Beam width

c Chamfer of a corner of a beam with 45° chamfers

d Beam depth

e Load eccentricity equal to $(a_1 - a)$

e/L Load eccentricity ratio equal to $(a_1 - a)/L$

e_c Shift of neutral axis in an initially curved beam

h_1, h_2 Horizontal shift of contact and load points due to beam bending

k_1, k_2 A numerical factor dependent upon b/d

ℓ Inner span length of a four-point loaded beam

ℓ' Equal to either a or $L/2$ for four-point or three-point beam systems

n Numerical factor

P_{max} Maximum contact pressure at the load application point

r Radius of the corners of the beam

s Speed of loading

t Time of loading

x_1, x_2, x_3 Variable beam distances

x' Variable distance (failure site location) on either side of the load contact point

x, y Coordinate axes

α_b, α_c Beam curvature parameters

$\dot{\epsilon}$ Strain rate

$\bar{\epsilon}$ Percent error, usually defined as $[(\sigma_b - \sigma_x)/\sigma_x]100$

μ Coefficient of friction

ν Poisson's ratio

ρ_1 Contact radius of a support point

ρ_2 Contact radius of a load point

ρ_c Initial radius of the curvature of a beam

σ_b Bending stress in a beam as defined by simple beam theory or mean fracture stress

σ_n Normal stress

$\sigma_{n_{max}}$ Maximum principal stress

σ_0 Scale parameter or characteristic value associated with a Weibull analysis

σ_x Stress in the x direction (along the beam length)

ϕ_s Angle of twist along the specimen length in radians

ϕ_F Angle of twist between a pair of load and contact points in relation to ϕ_s in radians

The flexure test is used extensively by ceramists to determine strengths of brittle materials. It is important to minimize errors that evolve from such testing. Obviously accuracy is important when the strength of a material is required for design applications and to ascertain structural reliability. Not so obvious is the importance of test accuracy when attempting to develop or improve a material. If the inaccuracy of a test system is greater than the relative difference in fracture strengths determined by a flexure test, then improvements will not be realized.

Flexure tests are generally assumed to yield accurate results and be easy to perform. These suppositions are usually correct if the beam specimen geometry satisfies simple beam theory assumptions and the loading fixtures are carefully designed; otherwise, excessive errors will result. A recent report [1] by this writer delineated in detail the errors that can arise in flexure testing of brittle materials at room temperature. Analyses have been documented [1], and for the sake of brevity the derivation details will not be included here. However, the report's results will be extensively referenced and used in this paper to describe, catagorize, and summarize those errors so that requirements for flexure testing of brittle materials at room temperature can be outlined. These results also provide guidance for a standard flexure test method [2] that has been developed.

Approach

Since the beginnings of the science of materials strength, attributed to Galileo [3] by Timoshenko [4], the beam as a structural element has been extensively analyzed [4]. Many of these applicable results, available from elasticity theory and strength of materials analyses, were used [1] and, where necessary, were modified or extended to problems pertinent to beam testing within the elastic region. Such sources of error fall into a natural classification: those that do not conform to simple beam theory assumptions, called here "internal error sources," and those arising from load application, termed "external error sources." Each of these error sources are, in turn, described and discussed in the paragraphs that follow.

Internal Error Sources

A critical review of beam theory assumptions will yield ranges of beam geometry ratios by which the theory can be applied validly. These assumptions are listed here, as well as their associated inferences in terms of an error analysis:

1. Transverse planes perpendicular to the longitudinal axis of the beam remain plane after the beam is bent.

2. The modulus of elasticity in tension is equal to the modulus of elasticity in compression; the beam material is isotropic and homogeneous.

3. The maximum deflection must be small compared to the beam depth.

4. The beam must deflect normally under elastic bending stresses but not through any local collapse or twisting.

Each of these assumptions is examined in detail, where possible, so that the required rectangular beam geometry ratios can be determined as a function of the associated errors.

Assumptions 1 and 2 together imply that stress and strain are proportional to the distance from the neutral axis, and the stress does not exceed the proportional limit of the material. These assumptions disregard the effect of any shearing resistance and make impossible the use of the flexure formula for curved beams of large curvature.

Assumption 1 and the previously stated implication suggest that the bending stress is proportional to the distance from the neutral axis to the outer surface of the beam. This assumption is valid if flexure of the beam could be attained without application of local forces to the beam. However, practical flexure test systems, such as those shown in Figs. 1a and 2a, which utilize four-point and three-point beams, require direct contact of the fixture with the specimen to apply loads, and thus moments, to the specimen. At the point of contact there will be compressive stress in the beam depth direction, resulting in a local variation from linearity in the bending stress [5]. Because this contribution to bending stress nonlinearity, referred to as wedging stress [6], is caused by external load application, it will be discussed in detail under the External Error Sources section.

An error source that is internal to the beam arises because of the assumption that the modulus of elasticity in tension is equal to that in compression, $E_T = E_c$. Chamis [7] has derived in closed form the solution for the tensile bending stress when $E_T \neq E_c$. After some manipulation of the appropriate formulas in Ref 7, the tensile stress due to bending is given by

$$\sigma_x = \frac{\sigma_b}{2}\left[1 + \left(\frac{E_T}{E_c}\right)^{1/2}\right] \tag{1}$$

for both the four-point and the three-point loaded beams. The resulting percent error is given in Table 1.

Although the errors associated with neglecting to account for anisotropy and nonhomogeneity of the test material are not considered here, they are briefly mentioned in the following paragraphs so that the reader will be aware of such possibilities.

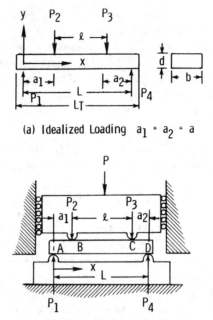

(a) Idealized Loading $a_1 = a_2 = a$

(b) Nonpivoting Rigid Loading Head
$a_1 \neq a_2$

FIG. 1—*Four-point loading.*

If the beam is anisotropic, the bending stress formula is exactly the same as the elementary theory except that the application of a bending moment produces a twisting moment, and vice versa. According to Lekhnitskii [8], determining the accompanying shear stress produced by bending a rod of rectangular cross section, having only one plane of elastic symmetry normal to the axis, is very complicated. (Composite and crystal structures are excluded here as test materials.) If the degree of anisotropy for a ceramic material is slight, it may be permissible to assume that the error incurred when ignoring this effect will also be small.

Nonhomogeneity of the test material implies variation of the elastic modulus. If the modulus variation is known, it can be treated in an analogous manner as an anisotropic beam, and thus the previous comments are applicable.

If a rectangular beam has initial curvature, ρ_c, the error can be determined from an analysis provided by Timoshenko [6]. The general bending stress σ_x in a curved beam due to a pure moment is given by the following:

$$\sigma_x = \alpha_c \left(\frac{M_b}{bd\rho_c} \right) \qquad (2)$$

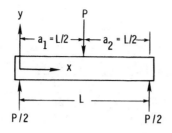

(a) Idealized Loading

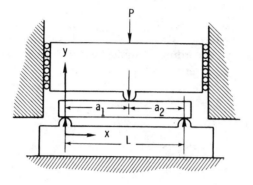

(b) Nonpivoting Rigid Loading Head; $a_1 \neq a_2 \neq L/2$

FIG. 2—*Three-point loading.*

TABLE 1—*Error when $E_T \neq E_c$.*

$\dfrac{E_T}{E_c}$	Percent Error	$\dfrac{E_T}{E_c}$	Percent Error
0.20	+38.2	1.025	−0.6
0.40	+22.5	1.050	−1.2
0.60	+12.7	1.075	−1.8
0.80	+5.6	1.10	−2.4
0.90	+2.6	1.15	−3.5
0.925	+1.9	1.20	−4.6
0.950	+1.3	1.30	−6.5
0.975	+0.6	1.40	−8.4
1.00	0	1.60	−11.7
		1.80	−14.6
		2.0	−17.2

where

$$\alpha_c = \frac{\dfrac{d}{2\rho_c} - \dfrac{e_c}{\rho_c}}{\dfrac{e_c}{\rho_c}\left(1 - \dfrac{d}{2\rho_c}\right)} \tag{2a}$$

and

$$\frac{e_c}{\rho_c} = \left[\frac{\left(\dfrac{d}{\rho_c}\right)^2}{12}\right]\left[1 + \frac{\left(\dfrac{d}{\rho_c}\right)^2}{15}\right] \tag{2b}$$

Since the bending stress, according to simple beam theory, is $\sigma_b = 6M_b/bd^2$, and putting σ_b in the same terms as Eq 2, we have

$$\sigma_b = \alpha_b\left(\frac{M_b}{bd\rho_c}\right) \tag{3}$$

where

$$\alpha_b = 6\left(\frac{\rho_c}{d}\right) \tag{4}$$

The percent error, $\bar{\epsilon}$, for a bent beam of rectangular cross section and of initial curvature to beam depth, ρ_c/d, resulting in a neutral axis shift of e_c/ρ_c is

$$\bar{\epsilon} = 100\left(\frac{\alpha_b - \alpha_c}{\alpha_c}\right) \tag{5}$$

The resulting error for a beam of rectangular cross section bent by a pure moment as obtained from Eq 5 is given in Table 2 as a function of initial curvature. It is assumed that an analogous analysis applied to a three-point loaded beam would produce similar results.

The validity of the assumption that the strain is proportional to the distance from the neutral axis is dependent upon the ratio of the beam width to its depth. Anticlastic curvature of rectangular beams or plates with intermediate ratios of b/d can lead to erroneous results using simple beam theory (see Timoshenko [9]). Of course, if the beam can be considered infinite in width, like a plate, the correction of the bending stress is simply $1/(1 - \nu^2)$ (see Baratta [10]). The question arises as to what ratios of b/d are appropriate for the application of simple beam theory. Ashwell [11] examined in detail the anticlastic curvature of rectangular beams and plates and provided the an-

TABLE 2—*Error caused by initial beam curvature.*

$\dfrac{\rho_c}{d}$	Percent Error [a]
1	35.1
2	16.7
3	10.9
4	8.4
10	3.2
15	2.2
20	1.7
40	0.8
100	0.3

$$\bar{\epsilon} = 100 \left[\frac{\alpha_b - \alpha_c}{\alpha_c} \right]$$

[a] All errors are negative.

swer to this question. The pertinent formulas taken from Ref *11* are given in Ref *1*. These equations were applied to ceramic materials with Poisson's ratio, ν, equal to 0.25, and the ratio of Young's modulus to fracture stress, E/σ_b, of 1000 to determine the percent error[2] using simple beam theory as a function of b/d, which is shown in Table 3.

If the maximum deflection is not small compared to the beam depth, linear beam theory cannot be employed without error. West [*12*] examined large deflections of three-point loaded beams, and from such results a definitive ratio of beam length to depth can be determined for valid application of simple beam formulas. Since for most brittle materials values of E/σ_b range approximately from 500 to 1000, the former value was used to compute the percent error because it would yield the largest error. Although the analysis was applied to a three-point loaded beam, the method was extended to determine errors for four-point loaded beams as well. The results of the calculations using the analysis of Ref *12* are presented in Table 4, which gives errors for the four-point and three-point loaded beam as a function of L/d.

It is implicit in the assumptions given in Ref *12* that the loads and moments are applied to the beam in an ideal manner with no friction occurring between the load application points and the beam. Ritter and Wilson [*13*] have determined a beam length-to-depth limit based on the minimization of friction effects when large deflections occur. The friction effect considered is that which gives rise to a moment caused by the slope at the load application point. Not considered in the analysis [*13*] are the effects of friction due to a moment acting out of the neutral plane of the beam, lateral contraction or

---

[2]Ashwell [*11*] considered a beam bent by a constant moment analagous to the four-point beam loading case, which represents a conservative bound on b/d for the three-point beam, as well.

TABLE 3—*Error caused by effect of anticlastic curvature ($E/\sigma_b = 1 \times 10^3$).*

$\dfrac{b}{d}$	Percent Error[a]
1.0	0
15.0	0
20.0	0.1
30.0	0.6
40.0	1.5
50.0	2.6
100.0	4.7
500.0	5.9
1000.0	6.1
∞	$\rightarrow (-\nu^2)100 = -6.25\%$

[a] All errors are negative.

extension, and changes in moment arms due to contact point tangency shift. These factors will be discussed later.

The results of Ref *13* provide an inequality for the four-point loaded beam which is given by the following limit equation to ensure negligible nonlinear deflections and friction effects:

$$\frac{\dfrac{L}{d} - \dfrac{a}{d}}{\dfrac{E}{\sigma_b}} \leq 0.3 \qquad (6)$$

The value of 0.3 was obtained from limiting the slope, as in Ref *13*, to less than 15° between the beam in the loaded and unloaded positions at the outer-

TABLE 4—*Error for beams with large deflection ($E/\sigma_b = 0.5 \times 10^3$).*

$\dfrac{L}{d}$	Percent Error Beam Loading[a]	
	Four-Point	Three-Point
0	0	0
25	0.1	0.1
50	0.6	0.4
100	1.4	1.0
150	2.5	1.8
200	4.1	2.4
250	7.0	4.9

[a] All errors are negative.

most support point. If the minimum value of E/σ_b is chosen to be 500, then we determine that for a four-point loaded beam Eq 6 becomes

$$\frac{L}{d} - \frac{a}{d} \leq 150 \tag{7}$$

It is noted from Table 4 that neglecting beam deflections resulted in greater error in calculation of bending stress for the four-point loaded beam than for the three-point loaded beam. To be conservative, therefore, it will be assumed that Eq 7 is applicable to the three-point loaded beam as well, with $a/d = 0$. Thus, Eq 7 becomes

$$\frac{L}{d} \leq 150 \tag{8}$$

It appears that these limits are compatible with those values given in Table 4, so that reasonable L/d ratios can be chosen that will result in small errors when deflection is minimized.

One of the last requirements, no buckling of the beam, is easily fulfilled for brittle materials with beam dimensions of practical test configurations. The reader can readily verify this statement by referring to Timoshenko and Gere [14].

Accuracy, which is implied in these restrictions, is also dependent upon the manner of load application, beam geometry, loading fixtures, and surface preparation. Although specimen size will not affect accuracy except for extremely small geometries, it will alter the magnitude of the stress level at failure, and this must also be considered. These subjects are discussed in the following paragraphs, and guidelines for specimen geometry and minimization of errors are provided.

First to be considered, however, are the merits of a four-point beam loading system in comparison with the three-point beam loading system.

Four-Point and Three-Point Loading

The bending moment, from which the desired fracture stress is computed in an idealized four-point beam loading system, as shown in Fig. 1a, is constant, and there are no horizontal or vertical shear stresses within the inner span. However, the bending moment in an idealized three-point beam loading system, shown in Fig. 2a, is linearly dependent upon the distance from the nearest support to the fracture origin and thus requires an additional distance measurement to accurately determine the fracture stress. Also, the shear stresses for the three-point beam loading system are developed over the full span, thus deviating from the ideally sought uniaxial stress state present in the four-point beam loading system.

Wedging stresses [6] are present under all points of load application during flexure testing of beams. The effect of the wedging stress occurring at the inner load points of a four-point beam test is to cause a deviation from the idealized calculated constant stress at the two local regions. However, if the ratio of half the distance between the outer span, L, and inner span, ℓ (called a), to beam depth d is great enough, the stress reduction will not only be small but will decay rapidly, and the stress predicted by simple beam theory will be developed. Yet, the maximum stress computed by simple beam formula for the three-point beam system is never attained. The actual maximum stress occurs at a short distance to either side of the center of the load application point, which can cause fracture at these sites, rather than at the center, according to Rudnick et al [15]. This observation has also been confirmed by Oh and Finnie [16], where only for a material with no scatter in strength will the fracture location of a three-point loaded beam be theoretically located at the central load point. However, in Ref 16 only a statistical analysis was considered.

Strength of brittle materials is a function of specimen size. Size effects can be accounted for through the use of statistical analysis offered by Weibull [17]. Although the four-point beam system ensures a simple stress state which is easier to analyze than the more complex biaxial stress state associated with the three-point beam specimen, this will be less of a consideration if the beam is designed properly. The three-point loaded beam system is preferred[3] in investigating material or process development because of smaller specimen size, or in attempting to pinpoint fracture origin location. On the other hand, the four-point loaded beam is preferred when determination of strength for design purposes is desired, because the center span is uniaxially stressed, that is, no shear stresses exist. It is concluded that each of these systems is suitable for a particular application and each has different advantages and disadvantages. Thus, both types of loading systems, the four-point beam system shown in Fig. 1 and the three-point beam system shown in Fig. 2, are considered as standards in Ref 2.

Each of these beam systems will be subjected to external influences which will affect the accuracy of the test results. These external influences, directly or indirectly caused by the application of loads through the test fixtures, will lead to either configuration constraints or errors.

External Error Sources

The major influence on the accurate determination of flexure strength of a beam in bending arises from the application of load through the fixtures to

---

[3]R. W. Rice, Naval Research Laboratory, personal communication, 1982.

the specimen. The idealizations indicated in Figs. 1a and 2a are rarely met, and often tests are conducted using a convenient rigid loading head and support member, as depicted in Figs. 1b and 2b. The constraints on either the loading fixture or the specimen and errors resulting from such fixture designs are many. Such constraints or errors, which are discussed in turn, are caused by the following:

(a) load mislocation,
(b) beam twisting,
(c) friction,
(d) local stresses,
(e) contact point tangency shift, and
(f) surface preparation.

Load Mislocation

Four-Point Loaded Beams—When calculating bending stress by simple beam theory formula for four-point loaded beams, it is usual to assume that the moment within the inner span is constant. However, if a loading head that can only translate is used, as idealized in Fig. 1b, it is impossible to attain this idealized moment condition when $a_1 \neq a_2 \neq a$ [15,18]; this is indicated in Fig. 1b. The ratio of σ_x/σ_b, from Ref 1 is

$$\frac{\sigma_x}{\sigma_b} = \left[\frac{P_1}{\dfrac{P_2 + P_3}{2}} \right] \frac{x_1}{a} \qquad (9)$$

The loads and distances are also shown in Fig. 1b, and a is the value of a_1 with perfect load location. The error is magnified by the ratio of x_1/a (of course, if $P_1 = P_2 = P_3$, which implies exact location of the points of load application, there is no error). In order to estimate the magnitude of such an error, it was assumed in Ref 1 that the upper two load points in Fig. 1b were at a fixed distance and were constrained to translate vertically during loading and that the loading head would be located so that $a_1 \neq a_2 \neq a$. The method of loading in which the loading head can only translate is often adopted by many investigators, and therefore the resulting error determination, although conservative, is not unrealistic.

The analysis was accomplished by simply enforcing the condition that the displacement at P_2 must be equal to the displacement at P_3 in the deflection equation. This results in the following relationships between σ_x and σ_b in terms of the load eccentricity ratio e/L:

$$\frac{\sigma_x}{\sigma_b} =$$

$$\frac{\left(\dfrac{\dfrac{e}{L}+\dfrac{a}{L}}{\dfrac{a}{L}}\right)\left[1-\left(\dfrac{e}{L}+\dfrac{a}{L}\right)-\dfrac{\ell}{L}\right]\left\{\dfrac{\ell}{L}\left[2-\left(\dfrac{e}{L}+\dfrac{a}{L}\right)\right]-2\left[1-\left(\dfrac{e}{L}+\dfrac{a}{L}\right)\right]^{2}\right\}}{3\left(\dfrac{e}{L}+\dfrac{a}{L}\right)\left[1-\dfrac{\ell}{L}-\left(\dfrac{e}{L}+\dfrac{a}{L}\right)\right]-\left(1-\dfrac{\ell}{L}\right)^{2}}$$

$$(10)$$

where the parameters, a, ℓ, and L, are shown in Fig. 1.

Most workers in the testing field utilize either a $1/3$-point ($a/L = 1/3$ and $\ell/L = 1/3$) or a $1/4$-point ($a/L = 1/4$ and $\ell/L = 1/2$) loading. Thus, by substitution of these parameters into Eq 10, we obtain

$$\left(\frac{\sigma_x}{\sigma_b}\right)_{\ell/L=1/3} = \frac{\left[3\left(\dfrac{e}{L}\right)+1\right]\left(\dfrac{1}{3}-\dfrac{e}{L}\right)\left[\dfrac{1}{3}\left(\dfrac{5}{3}-\dfrac{e}{L}\right)-2\left(\dfrac{2}{3}-\dfrac{e}{L}\right)^{2}\right]}{\left[3\left(\dfrac{e}{L}\right)+1\right]\left(\dfrac{1}{3}-\dfrac{e}{L}\right)-\dfrac{4}{9}}$$

$$(11)$$

and

$$\left(\frac{\sigma_x}{\sigma_b}\right)_{\ell/L=1/2} = \frac{\left[4\left(\dfrac{e}{L}\right)+1\right]\left(\dfrac{1}{4}-\dfrac{e}{L}\right)\left[\dfrac{1}{2}\left(\dfrac{7}{4}-\dfrac{e}{L}\right)-2\left(\dfrac{3}{4}-\dfrac{e}{L}\right)^{2}\right]}{\left[3\left(\dfrac{e}{L}\right)+\dfrac{3}{4}\right]\left(\dfrac{1}{4}-\dfrac{e}{L}\right)-\dfrac{1}{4}}$$

$$(12)$$

The reader is cautioned that for given values of e/L there exists a limit on e/L in Eqs 10, 11, and 12; that is, a_1 can be such that either P_2 or $P_3 = 0$ because the test system changes from four-point to an eccentric three-point loading.

The error, defined as $\{(\sigma_b - \sigma_x)/\sigma_x\}100$, was determined from Eqs 11 and 12 for the $1/3$-point and $1/4$-point loaded beams and is shown in Tables 5 and 6 as a function of e/L. Tables 5 and 6 show that for corresponding e/L, when $a_1/L \neq a_2/L$, the $1/3$-point loading system results in slightly less error than the $1/4$-point loading system. Also, in accordance with the preceding discussion, e/L in Tables 5 and 6 is limited to a range of 0.0443 to 0.0465. The errors indicated in these tables can be minimized by designing the loading fixture so that the inner and outer spans are independently fixed. Also, the inner span

TABLE 5—*Error due to eccentric load application for a $\frac{1}{3}$-four-point loaded beam (when $\ell/L = \frac{1}{3}$ and $a_1/L \neq a_2/L$).*

$\dfrac{e}{L} = \pm\left(\dfrac{a_1}{L} - \dfrac{1}{3}\right)$	$\pm$ Percent Error
0	0
0.0019	1.3
0.0038	2.6
0.0057	3.8
0.0076	4.9
0.0095	6.0
0.0114	7.0
0.0133	8.1
0.0333	16.1
0.0433	18.7
0.0443	18.9

should be designed with accurate location adjustment and allowed to pivot as recommended by Hoagland et al [18].

Three-Point Loaded Beams—The same type of analysis used to derive Eq 10 was applied to the three-point loaded beam. The ideal system is shown in Fig. 2a, and the system analyzed is shown in Fig. 2b. The resulting percent error due to an eccentric load application is given by

$$\bar{\epsilon} = \left[\frac{1}{4}\left(\frac{1}{1 - \dfrac{a_1}{L}}\right)\left(\frac{1}{\dfrac{a_1}{L}}\right) - 1 \right] 100 \tag{13}$$

The percent error as a function of e/L is given in Table 7.

TABLE 6—*Error due to eccentric load application for a $\frac{1}{4}$-four-point loaded beam (when $\ell/L = \frac{1}{2}$ and $a_1/L \neq a_2/L$).*

$\dfrac{e}{L} = \pm\left(\dfrac{a_1}{L} - \dfrac{1}{4}\right)$	$\pm$ Percent Error
0	0
0.0040	3.8
0.0080	7.1
0.0120	10.0
0.0160	12.6
0.0200	14.7
0.0240	16.6
0.0280	18.3
0.0320	19.8
0.0340	20.8
0.0400	21.8
0.0465	22.9

TABLE 7—*Error due to eccentric load application for a three-point loaded beam (when $a_1/L \neq a_2/L \neq \frac{1}{2}$ e/L = $\frac{1}{2}$ − a_1/L).*

$\pm\dfrac{e}{L}$	$\pm$ Percent Error [a]
0	0
0.025	0.25
0.050	1.0
0.075	2.3
0.100	4.2
0.150	9.9
0.200	19.0
0.250	33.3
0.300	56.3
0.400	177.8
0.450	426.3
0.500	∞

[a] All errors are positive.

Notice that the percent errors given in Table 7 are much less than those of equivalent e/L values shown in Tables 5 and 6 for the four-point beams. Notice also that when $\pm e/L = 0.500$, the error is infinite, that is, the three-point loading is no longer valid.

Beam Twisting

A net torque can result from line loads being nonuniform or nonparallel between pairs of load contact points or if the cross section of the specimen is skewed over its length [15,18]. This is referred to as a skewed system and is shown schematically in Fig. 3 for a four-point bending specimen. The error due to twisting has been estimated [18] for both the plane stress and plane strain conditions by examining the maximum principal stress due to bending and torsion, and then comparing it with the bending stress. Only the details

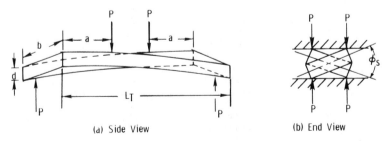

(a) Side View   (b) End View

FIG. 3—*Twisting of a four-point beam specimen.*

of the plane stress analysis were given in Ref *18*, and "bottoming" of the specimen on the fixture was not considered. Bottoming occurs in a skewed system when the load or support rollers contact the specimen across its full width.

The maximum principal stress for either a skewed four-point or three-point beam in bending, considering a plane strain condition, is given by

$$\sigma_{n_{max}} = \frac{\sigma_b}{2}\left\{1 + \nu + \frac{1}{3k_2}\left[\left(\frac{nb}{\ell'}\right)^2 + 9k_2^2(1-\nu)^2\right]^{1/2}\right\} \quad (14a)$$

where σ_b is the apparent bend strength and ℓ' is equal to either a for four-point bending or $L/2$ for three-point bending. Also

$$n = \left[\frac{3k_1\left(\frac{E}{\sigma_b}\right)}{1+\nu}\right]\left[\left(\frac{d}{L_T}\right)\phi_s + \left(\frac{d}{\ell'}\right)\phi_F\right]\frac{\ell'}{b} \quad (14b)$$

where for Case 1: $n = 1$, failure occurs prior to bottoming of the specimen in the loading fixture, and for Case 2: $n < 1$, failure occurs after bottoming.

The factors k_1 and k_2, obtained from Ref *5* and given in Table 8, are numerical values associated with the torsional stress component which are dependent on the ratio of b/d. The measured angle of twist (or skew angle) along the total length, L_T, of the specimen is ϕ_s (see Fig. 3), and along the fixture from one support point to the adjacent load point is ϕ_F.

The maximum principal stress, as given by Eq 14a, can be utilized to determine the percent error for various ratios of n, ℓ'/b, and b/d. This was accomplished and is shown in Table 9. Notice that the range of n varies from 0 to 1.00. It is expected that if bottoming does not occur prior to fracture because of an excessive twist angle, the maximum ratio of n that can be attained is 1.0, and thus the tables do not accommodate $n > 1.0$.

TABLE 8—k_1 and k_2.

$\dfrac{b}{d}$	k_1	k_2
1.0	0.1406	0.208
1.2	0.166	0.219
1.5	0.196	0.231
2.0	0.229	0.246
2.5	0.249	0.258
5.0	0.291	0.291
10.0	0.312	0.312
∞	0.333	0.333

TABLE 9—*Percent error due to beam twisting (v = 0.25).*[a]

	$\dfrac{\ell'}{b}$	$\dfrac{b}{d}$							
		1.00	1.2	1.5	2.0	2.5	5.0	10.0	∞
$n = 0.20$	1.0	3.18	2.88	2.61	2.32	2.12	1.68	1.47	1.30
	2.0	0.84	0.76	0.68	0.60	0.55	0.43	0.38	0.33
	2.5	0.54	0.49	0.44	0.39	0.35	0.38	0.24	0.21
	5.0	0.14	0.12	0.11	0.10	0.09	0.07	0.06	0.05
	10.0	0.03	0.03	0.03	0.02	0.02	0.02	0.02	0.10
	∞	0	0	0	0	0	0	0	0
$n = 0.40$	1.0	10.58	9.77	8.94	8.06	7.44	6.05	5.36	4.77
	2.0	3.18	2.89	2.61	2.32	2.12	1.68	1.47	1.30
	2.5	2.09	1.89	1.71	1.51	1.38	1.09	0.95	0.84
	5.0	0.54	0.49	0.44	0.39	0.35	0.28	0.24	0.21
	10.0	0.14	0.12	0.11	0.10	0.09	0.07	0.06	0.05
	∞	0	0	0	0	0	0	0	0
$n = 0.60$	1.0	19.01	17.75	16.51	15.11	14.11	11.79	10.58	9.54
	2.0	6.58	6.02	5.48	4.90	4.50	3.61	3.18	2.81
	2.5	4.44	4.04	3.67	3.26	2.99	2.38	2.09	1.84
	5.0	1.20	1.08	0.98	0.86	0.77	0.62	0.54	0.48
	10.0	0.31	0.28	0.25	0.22	0.20	0.16	0.14	0.12
	∞	0	0	0	0	0	0	0	0
$n = 0.80$	1.0	26.88	25.38	23.86	22.12	20.86	17.92	16.22	14.79
	2.0	10.58	9.75	8.94	8.06	7.44	6.09	5.36	4.77
	2.5	7.35	6.73	6.14	5.50	5.05	4.09	3.58	3.17
	5.0	2.09	1.89	1.71	1.51	1.38	1.10	0.95	0.84
	10.0	0.54	0.49	0.44	0.39	0.35	0.28	0.24	0.21
	∞	0	0	0	0	0	0	0	0
$n = 1.00$	1.0	33.75	32.13	30.47	28.54	27.11	23.64	21.73	20.03
	2.0	14.80	13.74	12.69	11.53	10.71	8.83	7.87	7.05
	2.5	10.58	9.75	8.94	8.06	7.44	6.05	5.36	4.77
	5.0	3.18	2.89	2.61	2.32	2.12	1.68	1.47	1.30
	10.0	0.84	0.76	0.68	0.60	0.55	0.43	0.38	0.33
	∞	0	0	0	0	0	0	0	0

[a] All errors are negative.

Friction

It has already been shown in Table 4, for the two beam systems considered, that the error due to deflection will be much less than $1/2\%$ if $L/d \le 25$. It appears that these limits are well within the geometry ratios required for the standards proposed in Ref 2. Therefore, the friction effect is minimized at the load and support points with regard to large deflections. This also implies that there will be no effect from friction on contact tangency shift. However, friction will cause a moment acting out of the plane of the beam that cannot be ignored. This factor is considered in the following.

When determining bend strength by simple beam theory, it is usual to assume that the supports and load points are frictionless, whereas in fact they are not. The presence of friction in flexure tests with fixed load and support points gives rise to couples at such locations as well as axial forces at the neutral axis of the beam. The net axial force is relatively small and therefore is ignored here. However, if the moment is not corrected to account for the couple in the determination of flexure stress, error will result. Error equations adapted from the results[4] available in the literature [18-20] are given here for the four-point and three-point loading systems.

$$\bar{\epsilon} = 100 \left(\frac{\mu}{\dfrac{a}{d} - \mu} \right) \tag{15}$$

and

$$\bar{\epsilon} = 100 \left(\frac{\mu}{\dfrac{L}{2d} - \mu} \right) \tag{16}$$

Such errors as defined by these equations can be significant, according to Refs 18, 20, and 21. Newnham [20] and Weil [21] reported that the experimental difference in failure stresses of silicon nitride and graphite using rigid knife edges compared with roller-type contact points was as high as 12% and 13%, respectively.

Local Stresses

Loads on bend specimens applied through knife edges or small-diameter rollers result in high compressive stresses, which can cause local crushing. Localized contact can also result in a more subtle problem, referred to as wedging stress.

Contact Stress—Given in Ref 6 are equations for determining the contact pressure between a cylinder (or roller) and a flat surface (see Fig. 4) as a function of the applied load, the modulus of each material, and the roller radius. If it can be assumed that the two materials are identical and that the allowable bearing pressure or contact pressure can be as high as twice the bend strength of the material, then limits on the roller radius for both loading systems will result:

$$p_{max} = 0.59 \sqrt{\frac{PE}{2b\rho_1}} \tag{17}$$

[4]Beam width constraint occurs also because of friction transverse to the beam's long axis. However, this effect [20] is small and thus is ignored.

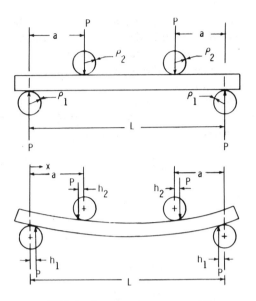

FIG. 4—*Contact point tangency shift.*

where p_{max} is the maximum contact pressure. (Note that the roller radius can be either ρ_1 or ρ_2.) We have assumed here that $p_{max} \leq 2\sigma_b$. Also for the four-point loading, $\sigma_b = 6Pa/db^2$, and for three-point loading, $\sigma_b = (3/2)PL/bd^2$. Substituting σ_b into Eq 17 and solving for ρ_1/d, we obtain

$$\frac{\rho_1}{d} \geq \frac{7.25}{\dfrac{a}{d}}, \text{ for the four-point loaded beam} \qquad (17a)$$

and

$$\frac{\rho_1}{d} \geq \frac{29.0}{\dfrac{L}{d}}, \text{ for the three-point loaded beam} \qquad (17b)$$

where it was assumed that $E/\sigma_b = 1 \times 10^3$.

Wedging Stress—The effect of wedging stress is to provide a substantial tensile stress contribution at the compressive side of the beam adjacent to the load points. A net tensile stress cannot be created if $d/2\ell' < 1$, according to Ref *18*. More important, a tensile stress is added to that already present at the tensile side of the beam, because of bending, thereby causing a deviation from the assumed stress calculated by simple beam theory.

This problem is generally treated in Ref *5*, and particular results from von

Kármán and Seewald [22] for a similar situation are used to estimate this error. A detailed analysis for this error is given in Ref 1. The resulting error determination for four-point and three-point loaded beams is given in Table 10. In the calculation of the errors, which are a function of a/d or L/d, as well as x'/d, the computed σ_b corresponds to the normalized failure site location (x'/d), where x' is the distance on either side of the load contact point.

Beam Overhang—The overhangs of the beam must be great enough that the local stresses at the beam support points are not amplified because of

TABLE 10—*Error due to wedging.*

				$\dfrac{x'^{a}}{d}$				
Loading	0	0.125	0.25	0.375	0.50	0.75	1.0	1.50
				FOUR-POINT				
$\dfrac{a}{d}$								
1.0	+4.7	−0.5	−2.8	−2.1	−1.4	−0.7	−0.3	0
1.5	+3.1	−0.3	−1.9	−1.4	−0.9	−0.5	−0.2	0
2.0	+2.3	−0.2	−1.4	−1.1	−0.7	−0.4	−0.2	0
3.0	+1.5	−0.2	−1.0	−0.7	−0.5	−0.2	−0.1	0
4.0	+1.1	−0.1	−0.7	−0.5	−0.3	−0.2	−0.1	0
5.0	+0.9	−0.1	−0.6	−0.4	−0.3	−0.1	0	0
6.0	+0.8	−0.1	−0.5	−0.4	−0.2	−0.1	0	0
8.0	+0.6	0	−0.4	−0.3	−0.2	−0.1	0	0
10.0	+0.4	0	−0.3	−0.2	−0.1	−0.1	0	0
15.0	+0.3	0	−0.2	−0.1	−0.1	0	0	0
20.0	+0.2	0	−0.1	−0.1	−0.1	0	0	0
40.0	+0.1	0	−0.1	−0.1	0	0	0	0
60.0	+0.1	0	0	0	0	0	0	0
∞	0	0	0	0	0	0	0	0
				THREE-POINT				
$\dfrac{L}{d}$								
1.0	+21.6	−2.4	−18.8	−25.4	−100.0	+6.2	+1.3	0
1.5	+13.4	−1.4	−10.4	−10.2	−10.1	−100.0	+2.6	0
2.0	+9.7	−1.0	−7.2	−6.4	−5.3	−5.5	−100.0	+0.1
3.0	+6.3	−0.7	−4.4	−3.7	−2.7	−1.9	−1.3	−100.0
4.0	+4.7	−0.5	−3.2	−2.6	−1.8	−1.2	−0.6	−0.1
5.0	+3.7	−0.4	−2.5	−2.0	−1.4	−0.8	−0.4	0
6.0	+3.1	−0.3	−2.1	−1.6	−1.1	−0.6	−0.3	0
8.0	+2.3	−0.2	−1.5	−1.2	−0.8	−0.4	−0.2	0
10.0	+1.8	−0.2	−1.2	−0.9	−0.6	−0.3	−0.2	0
15.0	+1.2	−0.1	−0.8	−0.6	−0.4	−0.2	−0.1	0
20.0	+0.9	−0.1	−0.6	−0.4	−0.3	−0.2	−0.1	0
40.0	+0.4	0	−0.3	−0.2	−0.1	−0.1	0	0
60.0	+0.3	0	−0.2	−0.1	−0.1	−0.1	0	0
∞	0	0	0	0	0	0	0	0

$^{a}x'$ is the distance on either side of the load contact point where failure occurs.

beam-end effects. These stresses are damped out within a distance equal to one beam depth, as determined from the data given in Ref 22. Thus, by allowing

$$L_T \geq L + 2d \tag{18}$$

we avoid beam-end effects.

Contact Point Tangency Shift

Significant changes in span length can occur in both four-point and three-point loading systems if contact radii of support and load points are large compared to beam depth. The shift in point of tangency, as shown by h_1 and h_2 in Fig. 4, is a function of the contact radii, specimen thickness, and the ratio of the modulus of elasticity to the bend strength. For materials that behave elastically, such as those considered here, the change in contact point location and thus the error arising because of the change in moment arm from the ideal can be predicted mathematically for linear systems. This has been accomplished in Ref 1. The approach was patterned after Westwater [23], who corrected for span shortening but ignored friction at the support points of a three-point loaded beam.[5]

In Ref 1 the formulas were derived for a four-point loaded beam and then reduced to the special case of a three-point loaded beam. These results were put in terms of error functions, assuming that simple beam theory was applied without correcting for the occurrence of span shortening in the lower support or span lengthening between the upper loading points, shown in Fig. 4.

The errors have been determined for $1/3$ and $1/4$ four-point loaded beams as a function of ρ_1/d and ρ_2/d, and the three-point loaded beam as a function of ρ_1/d only. It was assumed that E/σ_b equaled 1000. These errors are given in Table 11.

Surface Preparation

The flexure strength of each brittle material is not only supersensitive to the final surface finish, because the maximum tensile stress occurs at the beam surface, but is also highly sensitive to prior finish history. For this reason it is impossible to specify an optimum surface finish procedure for all brittle materials, so that failure will be due to inherent flaws related to the material or material processing, rather than to an imposed defect resulting from the fin-

---

[5]Westwater [23] also determined an approximate relationship for the horizontal load arising because of tangency shift. However, for beams of small deflection, the error is negligible.

TABLE 11—*Percent error due to tangency point shift* $(E/\sigma_b = 1 \times 10^3)$.[a]

Loading	$\dfrac{\rho_1}{d}$	$\dfrac{\rho_2}{d}$			
		1.0	2.0	5.0	10.0
Four-point,	1.0	0.3	0.4	0.7	1.2
$a/L = 1/3$	2.0	0.5	0.6	0.9	1.4
	4.1	0.9	1.0	1.3	1.8
	6.1	1.3	1.4	1.7	2.3
	8.2	1.7	1.8	2.1	2.7
	10.3	2.1	2.2	2.6	3.1
Four-point,	0.67	0.4	0.6	1.2	2.2
$a/L = 1/4$	1.35	0.6	0.8	1.4	2.5
	2.7	1.0	1.2	1.8	2.9
	4.1	1.4	1.6	2.2	3.3
	5.5	1.8	2.0	2.6	3.7
	6.9	2.2	2.4	3.1	4.1
Three-point,	1.0	0.1			
$a/L = 1/2$	2.0	0.2			
	4.0	0.4	regardless of ρ_2/d value		
	6.0	0.6			
	8.0	0.8			
	10.0	1.0			

[a] All errors are positive.

ish process (see Ref 2 for further discussion). Indeed, the designer or materials developer may not be able to specify a particular finish procedure. Therefore, rather than attempt to dictate surface finish requirements, it is suggested that each set of reported flexure test data results be accompanied by surface finish history or material process history, or both, whichever is applicable.

There are, however, several specific recommendations related to surface finishing procedures that can be presented. Corner flaws resulting from chipping or cracking during the grinding operation are sources of low-strength failure. Rounding or beveling of the corner, as depicted in Fig. 5, appears to reduce premature failure according to Rice [24]. Since a chamfer will double the number of edges, thus doubling the source of flaw locations, rounding is preferred [25]. Also, it is important to grind the edges and flat surfaces [25] by a motion parallel to, rather than perpendicular to, the specimen length. It is further indicated [24] that the finishing of the corner should be comparable in all respects to that applied to the beam surfaces.

If the corner radii or chamfer is small, the error in ignoring the change in moment of inertia will be negligible. The error as a function of the r/d or c/d ratio of corner radii or for a given b/d is indicated in Table 12.

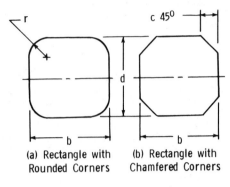

(a) Rectangle with   (b) Rectangle with
Rounded Corners   Chamfered Corners

FIG. 5—*Beam cross section.*

TABLE 12—*Percent error in determining flexure stress.*[a]

		$\dfrac{b}{d}$		
		1.0	2.0	4.0
	r/d			
When neglecting cor- ner radii	0	0	0	0
	0.02	0.1	0.1	0
	0.04	0.4	0.2	0.1
	0.06	0.9	0.4	0.2
	0.08	1.5	0.8	0.4
	0.10	2.4	1.2	0.6
	0.15	5.4	2.6	1.3
	0.20	9.7	4.6	2.2
	c/d			
When neglecting 45° chamfer	0	0	0	0
	0.01	0.1	0.1	0.1
	0.02	0.2	0.1	0.1
	0.03	0.5	0.3	0.1
	0.04	0.9	0.5	0.2
	0.05	1.4	0.7	0.4
	0.06	2.0	1.0	0.5
	0.08	3.4	1.7	0.9
	0.10	5.2	2.6	1.3

[a] All errors are negative.

Weibull Parameter Estimate and Sample Size

The type of analysis that has been used with varying degrees of success to relate failure strengths of brittle materials is attributed to Weibull [17]. Many investigators have used this approach to relate strength levels of various types of specimen configurations either on a stressed volume or surface area basis [26]. The reader is cautioned that confirmation of such an analysis or lack thereof may well depend on a number of factors, including the test material. As examples of such correlation and lack of it, Weibull statistical correlation was justified [26] for reaction-bonded silicon nitride but inappropriate, as reported by Lewis [27], in alumina fabricated by several processes.

A computer program for statistical evaluation of composite materials is available in Ref 28. This program determines the desirability of a particular probability density function in predicting fracture strength of ranked empirical data. The candidate functions include normal, log normal, and Weibull. Root mean square error results can be tabulated for each function comparison. The effects of statistical ranking can be readily listed in the computer output. The data mean and standard deviations with corresponding levels of confidence can be included in the printed results. The Weibull parameters, obtained from the maximum likelihood method, and corresponding confidence intervals can be obtained from this program [28].

Since a Weibull-type analysis is applicable in many instances, a discussion of Weibull parameter estimates and sample size determination is appropriate and given in the paragraphs to follow.

Different techniques will produce somewhat different results, according to McLean and Fisher [29], when estimating the Weibull parameters. Two statistical methods had been used during preliminary analysis of hot-pressed silicon nitride material strength data, and the results indicated that the estimates of the characteristic value σ_0 (or scale parameter) were very close whereas the Weibull slope estimates vary and thus would yield considerable differences in the component strength requirement.

The following is quoted directly from Ref 29 (except for changed reference and figure numbers as appropriate for this paper), because it succinctly addresses the question of proper sample size: "The exact confidence intervals for the parameters are based on the distributions obtained by Monte Carlo methods presented in Thoman et al [30]. It is not unexpected that the uncertainty in the estimation of a parameter will increase as the sample size decreases. This uncertainty, however, has rarely been quantified. The width of the confidence intervals for the parameters is a measure of the uncertainty and aids in the selection of the sample size of a test." Figures 6 and 7 are drawn from Ref 30 "and show the 90% confidence bounds for the Weibull slope and the characteristic value." (Figure 7 differs from that given in Ref 29 in that two additional M values were computed and shown.) "The bounds for the Weibull slope are a function of sample size only, while for the characteris-

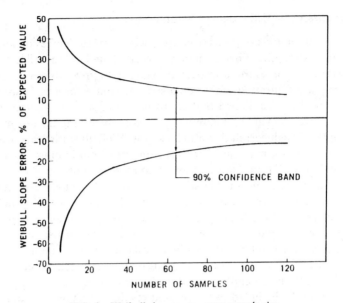

FIG. 6—*Weibull slope error versus sample size.*

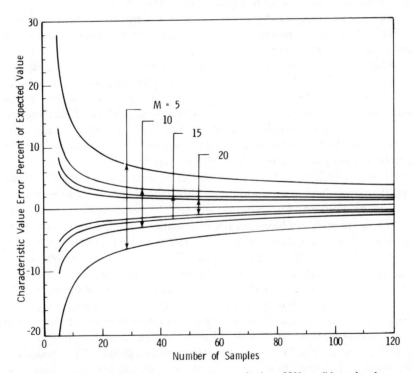

FIG. 7—*Characteristic value error versus sample size—90% confidence bands.*

tic value they are a function of both the sample size and the Weibull slope. As can be seen from the graphs, the error or uncertainty in estimates from small sample sizes is very large. Important judgments and significant analysis should not be based on small samples. Sample sizes of at least 30 should be used for all but the most preliminary investigations. An uncertainty of ±10% in Weibull slope requires more than 120 samples. This uncertainty is not peculiar to just ceramics, but is intrinsic to the statistical analysis of data, whether that data be material strength or the life of some electronic component. The choice of sample size depends on many factors including cost and time of testing and the degree of conservatism which is acceptable, but erroneous judgments may be made and unacceptable designs pursued if the sample sizes are too small."

Loading Speed

It is well known that brittle materials are strain-rate and environment sensitive, and thus speed of loading will influence the stress at which failure of the beam occurs. To choose a "static" speed, which would ensure no strain-rate effect, would increase the susceptibility of some materials to the effect of environment on fracture strength, and, conversely, a high rate of loading will cause a reverse trend. Nevertheless, most materials testing facilities utilize testing machines that allow a choice of test speeds, for example, 0.2, 0.5, and 1.0 mm/min. Therefore, it is possible to recommend one reference strain rate dependent upon the specimen sizes, and the following formulation for the strain rate $\dot{\epsilon}$ allowed this objective to be realized in Ref 2 for different sizes of four-point and three-point loaded beams:

$$\dot{\epsilon} = \frac{\dfrac{\sigma_b}{E}}{t} \tag{19}$$

where t is the time of the applied load; but since the speed of loading is $s = y/t$, then

$$\dot{\epsilon} = \frac{\dfrac{\sigma_b}{E}}{\dfrac{y}{s}} \tag{20}$$

where y is the deflection of the beam and s is the constant speed of the testing machine.

The deflection at the load points of a four-point loaded beam is

$$y = Pa^2 \frac{3L - 4a}{6EI} \tag{21}$$

Recalling that $Pa = 2\sigma_b I/d$, and substituting this and Eq 21 into Eq 20 gives

$$\dot{\epsilon} = \frac{3s}{\dfrac{a}{d}(3L - 4a)} \tag{22}$$

for the four-point loaded beam. Using the same approach as in Eq 22, we obtain the strain rate for the three-point loaded beam as

$$\dot{\epsilon} = \left(\frac{6d}{L^2}\right)s \tag{23}$$

Summary

A range of beam geometry ratios can be determined from previously presented tables and appropriate formulas if an accumulated error span of a given magnitude is chosen. For example, by examining those individual errors listed in the appropriate tables and assigning a total error composed of error limits from each source, limit bands on b/d, a/d, L/d, and ρ_c/d can be established for the four-point and three-point beam specimens [1,2]. Also, the limits on roller radius-to-beam depth ratios can be established [1,2] so that roller tangency point shift is not excessive and fracture does not occur at the contact points.

Those factors that arise from external error sources such as load mislocation, beam twisting, and friction at the load and support points are not necessary for arriving at the desired specimen geometry range. Nevertheless, these factors, resulting from load-fixture design, can be determined from the appropriate tables and formulas given in this paper and must be considered on an individual basis [1] in assessing the total error within a test system. These factors are usually responsible for major error contributions in attempts to determine the fracture strength of brittle materials utilizing a "simple" beam specimen.

Acknowledgments

The author wishes to acknowledge the constructive suggestions offered by G. D. Quinn and W. T. Matthews; both are members of the Army Materials and Mechanics Research Center.

References

[1] Baratta, F. I., "Requirements for Flexure Testing of Brittle Materials," AMMRC TR 82-20, The Army Materials and Mechanics Research Center, Watertown, Mass., 1982.

[2] Military Standard MIL-STD 1942 MR, *Flexural Strength of High Performance Ceramics at Ambient Temperature*, Army Materials and Mechanics Research Center, Watertown, Mass., January 1984.

[3] Galileo, *Two New Sciences*, Henry Crew and Alfonso de Salvio, Trans. Northwestern University, Chicago, Ill., 1939.

[4] Timoshenko, S. P., *History of Strength of Materials*, McGraw-Hill, New York, 1953.

[5] Timoshenko, S. and Goodier, J. N., *Theory of Elasticity*, 2nd ed., McGraw-Hill, New York, 1951.

[6] Timoshenko, S., *Strength of Materials: Part I—Elementary Theory and Problems*, 3 Vols., Van Nostrand, New York, 1958; *Part II—Advanced Theory and Problems*, 2nd ed., Van Nostrand, New York, 1942.

[7] Chamlis, C. C., "Analysis of Three-Point Bend Test for Materials with Unequal Tension and Compressive Properties," NASA TN D7572, National Aeronautics and Space Administration, Lewis Center, Cleveland, Ohio, March 1974.

[8] Lekhnitskii, S. G., *Theory of Elasticity of an Anistropic Elastic Body*, Holden-Day Series in Mathematical Physics, J. L. Bradstatter, Ed., Holden-Day, San Francisco, Calif., 1963, p. 204.

[9] Timoshenko, S., "Letter to the Editor," *Mechanical Engineering*, Vol. 45, No. 4, April 1923, pp. 259-260.

[10] Baratta, F. I., "When Is a Beam a Plate?" *Journal of the American Ceramics Society*, Vol. 64, No. 5, 1981, p. C-86.

[11] Ashwell, D. G., "The Anticlastic Curvature of Rectangular Beams and Plates," *Journal of the Royal Aeronautical Society*, Vol. 54, 1950, pp. 708-715.

[12] West, D. C., "Flexure Testing of Plastics," *Experimental Mechanics*, Vol. 21, No. 2, July 1964, pp. 185-190.

[13] Ritter, J. E. and Wilson, W. R. D., "Friction Effects in Four-Point Bending," *ASLE Transactions*, Vol. 18, No. 2, pp. 130-134, presented at the 29th Annual Meeting of the American Society of Lubricating Engineers, 28 April-2 May 1974.

[14] Timoshenko, S. and Gere, J. M., *Theory of Elastic Stability*, 2nd ed., McGraw-Hill, New York, 1961.

[15] Rudnick, H., Marschall, C. W., Duckworth, W. H., and Enrick, B. R., "The Evaluation and Interpretation of Mechanical Properties of Brittle Materials," AFML TR 67-316, U.S. Air Force Materials Laboratory, Wright-Patterson Air Base, Ohio, April 1968.

[16] Oh, H. L. and Finnie, I., "On the Location of Fracture in Brittle Solids—I, Due to Static Loading," *International Journal of Fracture Mechanics*, Vol. 6, No. 3, September 1970, pp. 287-300.

[17] Weibull, W., "Statistical Theory of Strength of Materials," *Royal Swedish Institute for Engineering Research*, Proc. No. 151, 1939, pp. 1-45.

[18] Hoagland, R. G., Marschall, C. W., and Duckworth, W. H., "Reduction of Errors in Ceramic Bend Tests," *Journal of the American Ceramics Society*, Vol. 59, Nos. 5-6, May-June 1976, pp. 189-192.

[19] Duckworth, W. H. Schwope, A. D., Salmassy, O. K., Carlson, R. L., and Schofield, H. Z., "Mechanical-Property Tests on Ceramic Bodies," WADC TR 52-67, Wright Air Development Center, Wright-Patterson Air Base, Ohio, March 1952, pp. 67-70.

[20] Newnham, R. C., "Strength Tests for Brittle Materials," *Proceedings of the British Ceramic Society*, No. 25, May 1975, pp. 281-293.

[21] Weil, N. A., "Studies of Brittle Behaviour of Ceramic Materials," TR 61-628, Part II, U.S. Air Force, Applied Sciences Division, Wright-Patterson Air Base, Ohio, April 1962, pp. 38-42.

[22] Von Kármán, T. and Seewald, F., Al.handl. aerodynam. Inst. Tech. Hochschule, Aachen, Vol. 7, 1927.

[23] Westwater, J. W., "Flexure Testing of Plastic Materials," *Proceedings*, Vol. 49, American Society for Testing and Materials, Philadelphia, 1949.

[24] Rice, R. W., "Machining of Ceramics," *Proceedings, of the Second Army Materials Technology Conference—Ceramics for High Performance Applications*, J. J. Burke, A. E. Gorum, and R. N. Katz, Eds., Brook Hill Publishing Co., Chestnut Hill, Mass., 1974.

[25] Rice, R. W., "The Effect of Grinding Direction on the Strength of Ceramics," *The Science of Ceramics Machining and Surface Finishing*, S. J. Schneider and R. W. Rice, Eds., SD Catalog No. 13.10, NBS Special Publication 348, U.S. Government Printing Office, Washington, D.C., 1972, pp. 365–376.

[26] Davies, D. G. S., "The Statistical Approach to Engineering Design in Ceramics," *Proceedings of the British Ceramics Society*, No. 22, 1973, pp. 429–452.

[27] Lewis, D., III, "An Experimental Test of Weibull Scaling Theory," *Journal of the American Ceramics Society*, Vol. 59, Nos. 11-12, 1976, pp. 507–510.

[28] Neal, D. M. and Spridigliozzi, L., "An Efficient Method for Determining the 'A' and 'B' Design Allowables," U.S. Army Research Organization Report 83-2, *Proceedings*, 28th Conference on the Design of Experiments in Army Research, Development and Testing, Monte Ray, Calif., 1983, pp. 199–235.

[29] McLean, A. F. and Fisher, E. A., "Brittle Materials Design, High Temperature Gas Turbine," Contract DAAG46-71-C-0162, Interim Report, AMMRC CTR 77-20, Ford Motor Company, Army Materials and Mechanics Research Center, Watertown, Mass., August 1977.

[30] Thoman, D. R., Bain, L. J., and Antle, C. E., "Inferences on the Parameters of Weibull Distribution," *Technometrics*, Vol. 11, 1969, pp. 445–460.

Summary

The purpose of the symposium on which this publication is based was to summarize the current state of knowledge on the use of fracture mechanics data in designing with brittle materials. To accomplish this purpose, papers were solicited that describe the analysis and testing of brittle materials and also the application of analytical and test results to actual structures. The symposium was successful in attracting twelve papers that describe the latest developments in these areas. A brief summary of each of these papers follows.

The paper by Marshall reports a study of acoustic wave scattering from flaws in both residual-stress and residual-stress-free fields. The acoustic scattering technique was shown to define flaw sizes accurately in the presence of residual stresses, which hold the flaws open. However, in residual-stress-free fields, the cracks tend to close and acoustic scattering techniques are not accurate. From the applications standpoint, machining flaws are generally located in residual-stress fields; consequently, their flaw sizes can be accurately determined and the fracture behavior of those materials predicted.

Cook and Lawn introduced indentation flaws of various sizes into a wide range of glass and ceramic materials by controlling the contact loads. These authors subsequently tested the indented specimens to failure under controlled stress and environmental conditions. The test results were used to generate dynamic-fatigue and static-fatigue master maps for these materials. These maps provide simple graphic illustrations of the relative toughness and fatigue resistance of the various materials.

Gonzalez et al describe a fatigue study on polycrystalline alumina specimens containing either processing-induced ("natural") or indentation-induced flaws. This study shows that the scatter in inert strength is smaller for indented specimens than for ones containing natural flaws. It further indicates that, for indented specimens, the inert strengths are substantially higher than the fatigue strengths, even at the highest loading rates. The authors used the test results on the indented specimens to predict the static fatigue behavior of the nonindented specimens. Generally, these predictions were quite good. It was further found that residual stresses play a critical role in the behavior of all flaws.

223

Shetty et al review the results of tests on alumina subjected to either uniaxial or biaxial stress states. The tests were conducted in inert and water environments. The results of the uniaxial tests in an inert environment, and of the uniaxial and biaxial tests in water, correlate quite well with statistical fracture theory. However, the strengths of the specimens in the biaxial tests in an inert environment are considerably higher than those predicted by theory. To explain this anomaly, the authors propose that the strength of alumina in water drops off much more rapidly under a biaxial stress state than under a uniaxial one.

Magida et al developed design diagrams for a machinable glass-ceramic material, Macor. (This material will be used for structural beams in an orbiting observatory.) The design diagrams were constructed using the results of fatigue and impact tests, fracture mechanics technology, and statistical analysis procedures. Proof-test stress levels were also developed for establishing initial flaw sizes in the material. Using (1) the design diagrams for the material and (2) the proof-test technique for initial flaw size identification, the allowable stress levels for the design life of the beams can be determined.

Wiederhorn et al used crack velocity–K_I curves to predict strength distributions for proof-tested soda lime silica glass specimens. They then experimentally determined the strength distributions for the proof-tested specimens and compared them with the predictions. They found generally poor correlations between the predicted and the experimentally determined strength distributions. They conclude that the strength distribution after proof-testing is extremely sensitive to the position and shape of the crack growth curve. Thus, any process which disturbs the crack growth curve will cause errors in the prediction of the subsequent strength distribution.

Trantina investigated the use of crack velocity and stress rupture data to predict initial flaw size distributions in soda lime glass. Both types of data were taken from independent studies reported in the technical literature. To make the predictions, a relationship was first fitted to the crack velocity data. This relationship was then numerically integrated from an initial flaw size to the critical flaw size in the stress rupture tests. This process was repeated for each data point from the stress rupture test. The resulting distribution of initial flaw sizes was analyzed using the two-parameter Weibull distribution function. This function provided a reasonably good fit to the size distribution of critical defects. It also demonstrated that the defect distribution was independent of the stress level. The author then proposes that the specimen size effect can be given in terms of the square root of the average initial defect size.

Ritter et al analyzed a compilation of both static and dynamic data for optical glass fibers. These data were generated at a wide range of temperature and humidity conditions. A stress corrosion model was used in analysis of these data. None of the data analyzed were found to agree with this model. The authors propose either that the polymeric coatings on the fibers induce irregular resistance to moisture or that the fatigue mechanisms are different at different stress levels. Based on these findings the authors recommend an empirical approach to lifetime predictions. They carefully point out, however, the dangers of extrapolating outside of the data base when making lifetime predictions.

Dabbs et al discuss two aspects of indentation flaw generation in ultrahigh-strength glass. They studied both the kinetics of crack initiation and the fatigue properties of specimens having subthreshold and postthreshold indentation flaws. The kinetics study showed that shear faults within the deformation zone act as crack initiation sites. Further, the time required to initiate cracks decreases as both the contact load and the water content of the environment increase. In the fatigue properties study, subthreshold flaws (1) failed at higher loading levels, (2) experienced greater reductions in strength in the presence of water, and (3) exhibited greater data scatter than did postthreshold flaws. As did Ritter et al, the authors recommend an empirical approach to lifetime predictions and recommend caution when extrapolating outside the available data.

Fett and Munz examined the fracture behavior of silicon nitride at elevated temperatures. Tests were conducted at both constant displacement rates and constant applied loads. Linear elastic fracture mechanics relationships, used in conjunction with the constant displacement rate test results, successfully predicted the results of the constant applied load tests. However, at the highest temperature levels, stress changes due to creep had to be accounted for in order to make accurate predictions.

Quinn shows that more than one mechanism can be active in the failure of high-performance ceramics, for example, silicon nitride and silicon carbide. For silicon nitride, slow crack growth tends to dominate if the stresses are higher and the temperatures are somewhat lower. Conversely, creep fracture appears to dominate if the stresses are lower and the temperatures higher. For silicon carbide, slow crack growth tends to dominate at temperatures above 1300°C, and an undefined, alternate mechanism dominates at lower temperatures. The author points out that most failure mechanisms assume a homogeneous material, whereas in many cases the materials are not homogeneous. Surface conditions, grain boundary conditions, and the basic structure of the material can vary widely. Consequently, the static fatigue behavior and the failure mechanism of the material can also vary.

Baratta presents a splendid review of potential sources of both internal and external errors in the testing of brittle materials. Internal error sources include (a) the assumption that the moduli of elasticity in tension and compression are equal, (b) material anisotropy and nonhomogeneity, (c) specimen curvature, both initial and anticlastic, (d) large specimen deflections, and (e) nonlinear deflections and some frictional effects. External error sources include load eccentricity, specimen twisting, wedging, local stresses, contact point shifts, and surface conditions. Guidelines for minimizing and accounting for these errors are provided. The advantages and disadvantages of the three-point and four-point bend specimens are also reviewed. This paper is an excellent review for scientists and engineers who are just beginning the study of fracture in brittle materials.

C. Michael Hudson

NASA Langley Research Center, Hampton, Va. 23665; symposium chairman and editor.

Stephen W. Freiman

National Bureau of Standards, Washington, D.C. 20234; symposium chairman and editor.

Index

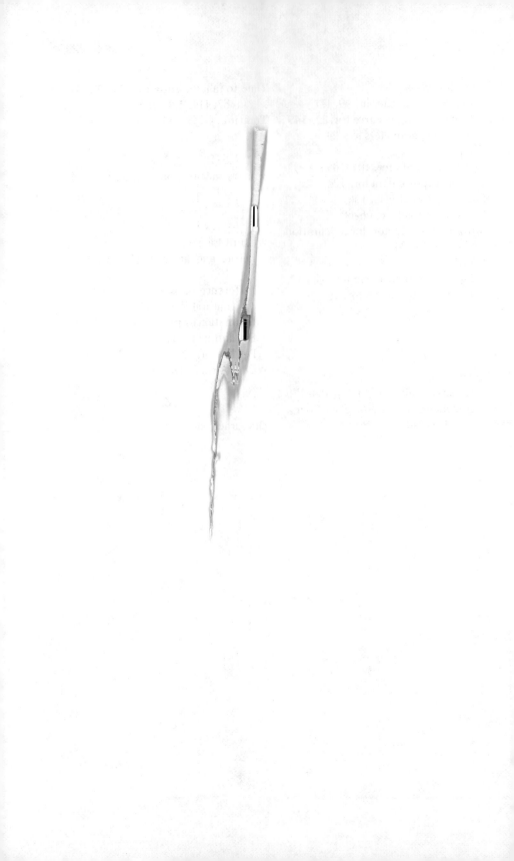